Analyzing Mathematical Patterns – Detection & Formulation

Inductive Approach to Recognition, Analysis and Formulations of Patterns

Analyzing Mathematical Patterns – Detection & Formulation

Inductive Approach to Recognition, Analysis and Formulations of Patterns

Michael A Radin
Rochester Institute of Technology, USA

NEW JERSEY • LONDON • SINGAPORE • BEIJING • SHANGHAI • HONG KONG • TAIPEI • CHENNAI • TOKYO

Published by

World Scientific Publishing Co. Pte. Ltd.
5 Toh Tuck Link, Singapore 596224
USA office: 27 Warren Street, Suite 401-402, Hackensack, NJ 07601
UK office: 57 Shelton Street, Covent Garden, London WC2H 9HE

Library of Congress Cataloging-in-Publication Data
Names: Radin, Michael A. (Michael Alexander), author.
Title: Analyzing mathematical patterns — detection & formulation :
inductive approach to recognition, analysis and formulations of patterns /
Michael A Radin, Rochester Institute of Technology, USA.
Description: New Jersey : World Scientific, [2023] | Includes bibliographical references and index.
Identifiers: LCCN 2022024371 | ISBN 9789811261046 (hardcover) |
ISBN 9789811262104 (paperback) | ISBN 9789811261053 (ebook for institutions) |
ISBN 9789811261060 (ebook for individuals)
Subjects: LCSH: Pattern perception--Mathematics. | Sequences (Mathematics) |
Pattern recognition systems--Mathematical models.
Classification: LCC Q327 .R33 2023 | DDC 516/.15--dc23/eng20220825
LC record available at https://lccn.loc.gov/2022024371

British Library Cataloguing-in-Publication Data
A catalogue record for this book is available from the British Library.

For any available supplementary material, please visit
https://www.worldscientific.com/worldscibooks/10.1142/12986#t=suppl

Desk Editors: Jayanthi Muthuswamy/Nijia Liu

Typeset by Stallion Press
Email: enquiries@stallionpress.com

Preface

Numerous patterns often emerge in artistic decorations and in architectural designs. Detecting patterns, examining their unique features and formulating them is essential for the development and enhancement of our intuitions, analytical skills and inductive and deductive reasoning skills. It is crucial to detect and formulate specific patterns while studying weather systems, learning to play a musical instrument, studying a foreign language, learning computer programming, studying traffic patterns and transportation schedules, analyzing engineering structures, processing signals, studying behaviors, studying economic cycles and additional analogous applications.

The goals of this interdisciplinary book are to gain practical experience in detecting the rise of specific patterns while studying complex structures of geometrical figures such as systems of squares, systems of triangles, systems of circles, systems of right triangles and additional geometrical systems. This will direct you to the discovery route to repetitions of patterns as same scale, repetitions of patterns at different scales and alternating patterns. These analyses will guide you to the examination of describing peculiar sequences such as linear sequences, quadratic sequences, summation-type sequences, geometric sequences, product-type sequences, factorial-type sequences, piecewise sequences, recursive sequences, periodic sequences and the characteristics of the Pascal's triangle.

I invite you to this discovery voyage in detecting, analyzing and formulating various patterns geometrically and analytically. The intricate study of patterns will help you nourish your intuitive skills and will guide you to an inductive approach and cognizance of other branches of mathematics and STEM related disciplines. For each topic, step by step repetitive-type examples will be provided. This will then direct you to inductively expanding your knowledge and comprehension about assorted patterns and their applications.

During the last 6 years, I successfully implemented the hands-on teaching style with repetitive-type guided examples in my courses that I teach at the Rochester Institute of Technology. In addition, I had the opportunities to implement these techniques in my courses and seminars that I teach annually in Latvia at Riga Technical University, at Rezekne Technical Academy and at Liepaja University and the Transportation and Sakaru Institute. Furthermore, I successfully implemented the use of various colors to emphasize the grouping of patterns and specific terms of a pattern while formulating piecewise sequences and developing and formulating proofs of theorems. The students often called it the "Four-color Technique" in their teaching evaluations. It is an honorable experience to receive positive and supportive students' and colleagues' feedback and evaluations and their constructive critiques that led to further ideas and discoveries.

This textbook can be used to teach courses such as discrete mathematics, introduction to patterns, introduction to recursive patterns, introduction to geometrical patterns, introduction to periodic patterns and prepares students for more challenging mathematics courses. The textbook's goal primary goal is to get students in Non-STEM disciplines acquainted with the fundamentals and applications of patterns. Furthermore, the aim of this textbook is to prepare students for future courses such as calculus, introduction to graph theory, introduction to difference equations, linear algebra, abstract algebra and other challenging courses.

Moreover, the aims of this textbook are to encourage professors and teachers to design and teach an introductory course to get students acquainted with pattern recognition. To encourage teachers and professors to use the textbook to teach various recreational mathematics, diversity of mathematics to prepare their students for further challenging mathematics and STEM-related courses.

Michael A Radin

About the Author

Michael A Radin is an associate professor of mathematics at the Rochester Institute of Technology, USA. Michael is an international scholar and enjoys developing and teaching mathematics and multidisciplinary courses and seminars in the traditional and in the online teaching and learning environments in different academic systems. Michael spent his spring 2009 sabbatical teaching at the Aegean University in Greece and his spring 2016 sabbatical teaching in Latvia. Michael gained vast international experiences while teaching in Greece, Latvia, Ukraine, Poland and Russia. Recently Michael developed and taught a course on "Pattern Recognition" for high school students in Rezekne, Latvia and developed and taught "Introduction to Business Start-Ups" in Ukraine and Poland. Michael considers teaching as a hobby and his primary aim is to make mathematics accessible to students with different preparation levels and from different disciplines in the traditional face-to-face and online environments. Michael's goals are to make online, international and multidisciplinary education accessible and to inspire students to learn. In closing, Michael enjoys exploring new teaching innovations in the American and foreign educational systems and comparing their similarities and contrasts and keep expanding the international and multidisciplinary frontiers.

Acknowledgments

First, I would like to take the opportunity to give special thanks my editor Rochelle Kronzek Miller for all her efforts, her support and guidance with this new textbook topic. Her encouragements certainly guided me in new innovative directions and to new experiences and intuitions. Her suggestions were very beneficial with the textbook's structure such as adding more detailed transient steps in guided examples, introduction of definitions, with the use of proper colors in examples that emphasized the decomposition of piecewise patterns and with detailed feedback on the graphics in the book. She taught me new techniques how to present the unique traits of each specific concept.

Second of all, I would like to thank the reviewers for their meticulous observations and suggestions while reviewing the book. Their vigilant comments very useful and enhanced the book's contents, transitions between topics and guided examples. Their prudent recommendations also directed me to new ideas for future textbooks.

I would like to thank my colleagues at the Rezekne Technical Academy High School in Rezekne, Latvia. Aivars Vilkaste (the school director), Vineta Pavlova (English teacher) and the students for their support with the first pilot mini-course on "Introduction to Recognition of Patterns and Deciphering of Patterns" that I conducted there in May 2019. Their supportive feedback led me to new valuable experiences and to new teaching innovations, practices and principles.

I would also like to thank my colleagues from the Transportation and Sakaru Institute in Riga, Latvia for giving me an opportunity to teach "Introduction to Discrete Mathematics" during my spring 2016 sabbatical in Latvia. This was a very beneficial teaching, cultural and international experience for me. This experience welcomed me to cross-cultural comparison and to new recognition of mathematical patterns. Students' active participation navigated me to alternative and non-standard thinking while solving specific problems. In particular, I would like to say special thanks to Irina Jackiva, Dmitri Pavluk, and Boriss Mishnevs for presenting these unique and beneficial international and cross-cultural experiences.

Moreover, I would like to thank Daniil Timchenko, a high school student from Riga, Latvia who identified several mistakes while he practiced working out the book's examples and the end of chapter exercises. Thanks to Daniil, the end of chapter exercises are more diverse and provide a broader range of challenge levels.

Finally, I would like to thank my parents Alexander and Shulamit for encouraging me to write textbooks, for their support with the textbook's content and for persuading me to continue writing future textbooks. In closing, they said "This will be one of many books ahead".

Contents

Chapter 1

Introduction to Patterns

You define a **pattern** as "repeated design or a recurring sequence, an ordered set of numbers, shapes or other mathematical objects arranged according to a rule". Recognition and formulation of patterns is an essential part of the learning process as you note patterns during specific instances such as: traffic patterns, musical patterns, languages' patterns, behavioral patterns in psychology, patterns in decision making, nature's formations patterns, weather patterns (winds, storms, waves, etc.), patterns in computer programs, signals' patterns and signal processing, decorative patterns, architectural patterns and mathematical patterns. You often notice patterns repeating at the same scale and at other instances repetition of patterns at different scales. In addition, you can discover alternating patterns. First let's begin analyzing some of nature's patterns. The consequent aerial photograph paints the repetitive alpine patterns of the Swiss Alps at the same scale:

You can detect similar nature's patterns at the same scale such as ocean's waves, dunes' formations, scattered clouds, archipelago of islands, etc. In Chapters 2 and 3, you will examine various geometrical arrangements and piecewise functions at the same scale. Next let's focus on repetition of patterns at different scales. The upcoming aerial photograph describes a system of repeated canyons at different scales:

You can discover similar nature's patterns at different scales such as tree's branches, leaf's roots, descending alpine crests, etc.

In Chapters 2 and 3, you will examine geometrical arrangements at different scales. Next let's direct your focus on alternating patterns. The succeeding aerial photograph resembles the clouds' alternating theme and their traced reflections in the Atlantic Ocean:

One of the main objective of the book is to get acquainted with when and how specific patterns come about. You will also learn how to identify the distinct patterns geometrically and analytically through the use of repetitive-type guided examples and with different colors. You will examine assorted patterns in numerous geometrical applications and will use the distinct patterns as an essential counting technique. Sometimes it will not be possible to write a formula for a specific pattern within one fragment alone. Instead, it will require a "piecewise pattern" that consists of at least two patterns or sub-systems. You will sometimes decompose a specific pattern into sub-patterns or into sub-systems of patterns with different colors.

Repetitive-type of guided examples will lead you to identifying the patterns' fundamentals and formulations. You will formulate patterns after numerous repetitions, when and how patterns emerge, their distinct traits and applications. You will begin to understand how patterns relate to each other. Once you are comfortable with detecting the patterns, you will be able to connect to the development and formulations of theorems and their proofs. Learning this essential technique will direct you to discovering the proof using the proof by induction method and formulating the correct result.

1.1 Geometrical Arrangements

Your journey begins with some numerical detection and formulation of patterns using several geometry examples that characterize decomposition of figures at the same scale. You will encounter **geometrical fractals** that resemble the decomposition of figures at diminishing scales.

Figure 1.1 examines a system of identical right triangles replicated at the same scale.

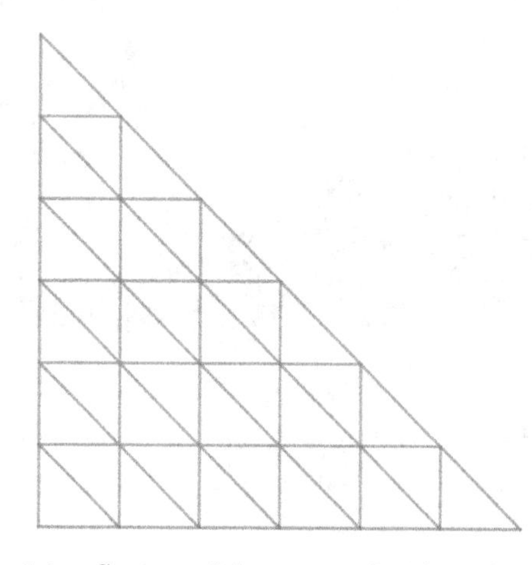

Figure 1.1: System of decomposed right sub-triangles.

First, note that Figure 1.1 decomposes the primary right triangle into smaller right triangles. Next, observe that Figure 1.1 has six rows of triangles and each row has an odd number of triangles. In fact, the very top row consists of one triangle, the second row has 3 triangles, the third row has 5 triangles and so on. In Chapter 2, you will pose the following question: how many sub-triangles are inside the primary triangle? This will require the knowledge of a specific pattern whose sum adds up to a perfect square. You will examine and formulate this pattern in Chapters 2 and 3. To sketch the assigned system of triangles in Figure 1.1, we will require sketching horizontal, vertical and diagonal lines.

Similarly, Figure 1.2 examines a system of right triangles with alternating green and blue colors together with alternating triangular shapes (triangles facing upwards and downwards).

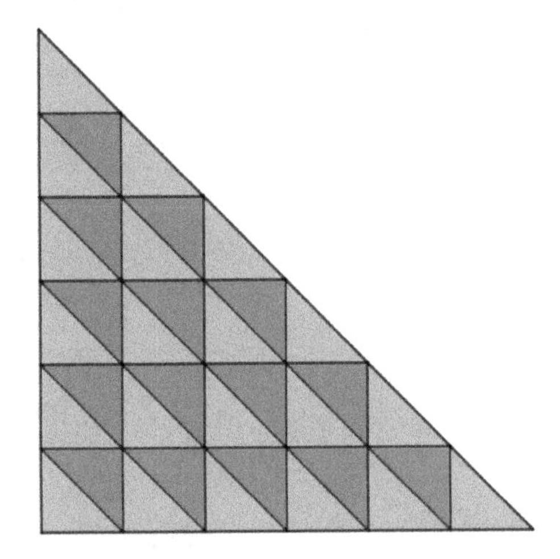

Figure 1.2: System of alternating green and blue right triangles.

First, note that the largest or principle equilateral triangle is decomposed into smaller equilateral triangles at the same scale with alternating green and blue colors and with alternating triangular shapes. Figure 1.2 is assembled with several right triangles with blue and green colors. Notice that any two neighboring triangles in the horizontal and vertical directions have different colors. Next, observe that any two neighboring triangles face in the opposite direction; that is, the green triangles face upwards while the blue triangles face downwards. The green and blue triangles are "sub-patterns" of the primary right triangle.

How many green triangles and how many blue triangles are inside the principle triangle in Figure 1.2? This question will be examined in Chapters 2 and 3, where you will formulate a specific pattern that adds consecutive positive integers that determine the exact number of triangles. This presents an alternating pattern from the alternate blue and green triangles. Next let's focus on repetitions of patterns at different scales.

What are geometrical fractals? You define a **fractal** as a repetition of patterns or shapes at different scales. Figure 1.3 analyzes a system of diminishing equilateral triangles. The dominant blue equilateral triangle contains smaller equilateral triangles produced by connecting the lines between the three midpoints of each equilateral triangle.

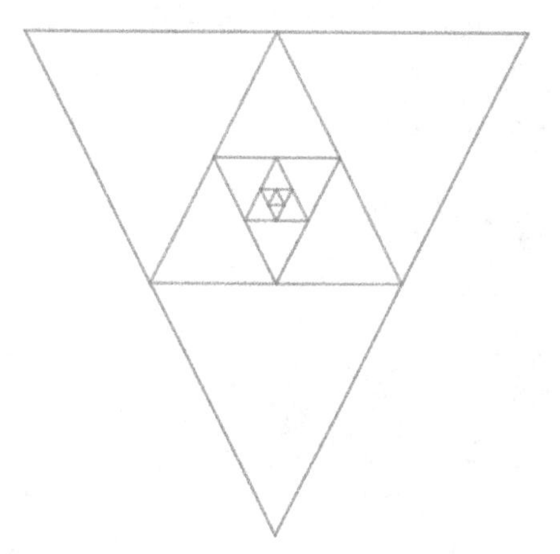

Figure 1.3: System of diminishing equilateral triangles.

Figure 1.3 is a system of shrinking triangles constructed by connecting the lines of the green triangle between the midpoints of the main blue triangle and so on. Observe that the green triangles and blue triangles decrease as opposite patterns. The blue triangles are facing downwards while the green triangles are facing upwards. In addition, the green triangle is assembled by reflecting the blue triangle vertically while it declines in size. How many triangles are inscribed inside the blue focal triangle? This question will be addressed in Chapters 2 and 3. To answer this question, you will need to formulate the process with a **linear sequence** or a **linear pattern**.

On the contrary, the dimensions of the triangles are described by a **geometric sequence** or a geometrical pattern. This also presents an **alternating sequence** or an alternating pattern as the shapes of the triangles alternate.

Figure 1.4 traces a system of diminishing right triangles embedded inside the principle blue right triangle.

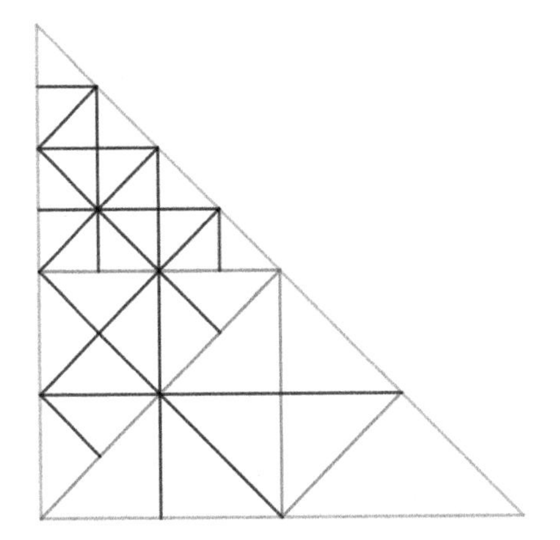

Figure 1.4: System of diminishing right triangles.

Note that in Figure 1.4, the principle blue right triangle contains groups of diminishing triangles starting with the largest triangle in the lower right corner. How will you be able to determine the total number of inner triangles? This question will be addressed in Chapter 3 where you will learn to use a specific **geometric summation** to determine the total number of inner triangles, not including the main blue right triangle.

You will examine more comprehensive examples of related patterns of geometrical systems in Chapters 2 and 3 and how to formulate these patterns analytically by using linear and geometric patterns.

1.2 Piecewise Functions

What patterns emerge as "piecewise functions" graphically? A **piecewise function** is a function that consists of two or more functions (fragments) on different restricted intervals. Piecewise functions naturally appear in numerous applications describing economic cycles, population dynamics, weather patterns, signal processing, neural networking, and other similar natural phenomena.

Figure 1.5 describes a sequence of positive integers as a system of horizontal lines on the corresponding restricted intervals with length 1.

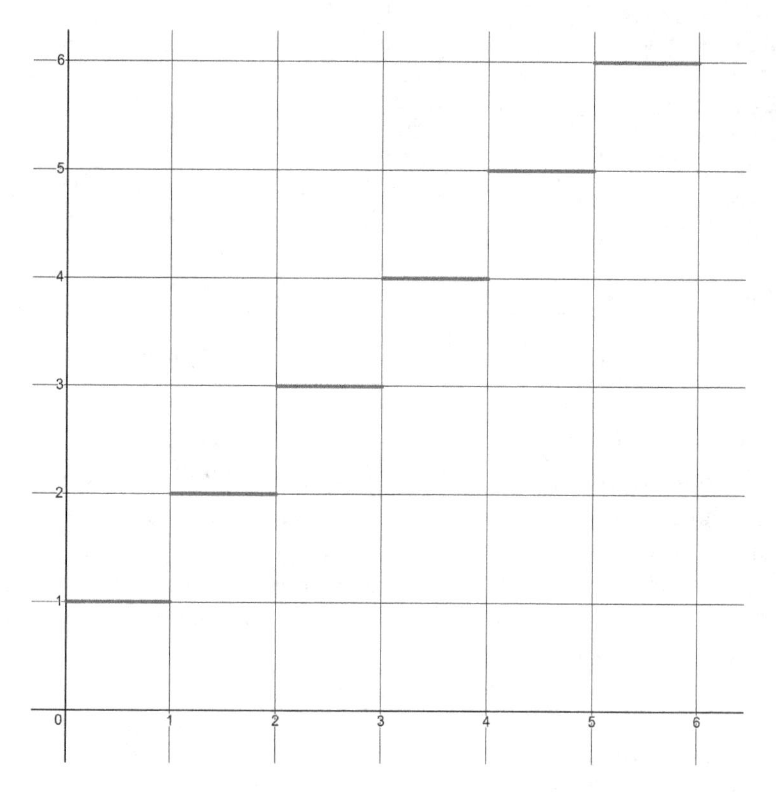

Figure 1.5: System of escalating horizontal lines.

The first horizontal line $y = 1$ is on the interval $(0, 1)$, then $y = 2$ is on the interval $(1, 2)$, then $y = 3$ is on the interval $(2, 3)$ and so on. Note that the horizontal lines escalate as steps in increments of one from one neighboring interval to the next starting with the interval $(0, 1)$, then shifting to then interval $(1, 2)$, then to the interval $(2, 3)$ and so on. The left end-points of the restricted intervals in Figure 1.5 are described by the following sequence:

$$0,\ 1,\ 2,\ 3,\ 4,\ \ldots. \tag{1.1}$$

The right end-points of the restricted intervals in Figure 1.5 emerge as the corresponding sequence:

$$1,\ 2,\ 3,\ 4,\ 5,\ \ldots. \tag{1.2}$$

Therefore, via (1.1) and (1.2), for all $n \in \mathbb{N}$ you formulate the cognate piecewise equation of horizontal lines presented in Figure 1.5:

$$y = \begin{cases} 1 & \text{if } \ x \in (0, 1) \\ 2 & \text{if } \ x \in (1, 2) \\ 3 & \text{if } \ x \in (2, 3) \\ 4 & \text{if } \ x \in (3, 4) \\ \vdots & \\ n & \text{if } \ x \in (n-1, n) \\ \vdots & \end{cases} \tag{1.3}$$

Furthermore, (1.3) lists all the consecutive positive integers (natural numbers $\mathbb{N}$) starting with 1 as:

$$1,\ 2,\ 3,\ 4,\ 5,\ 6,\ \ldots. \tag{1.4}$$

You encountered (1.4) in Figure 1.2. You will analyze deeper specifics of (1.4) and analogous patterns in Section 1.3 and in later chapters.

Analogous graphs of piecewise functions can be convened with their corresponding patterns and will be left as end of chapter exercises in this chapter and in Chapter 2.

Figure 1.6 describes an oscillatory and alternating character between the horizontal lines $y = 1$ and $y = -1$ on corresponding restricted intervals with length 1.

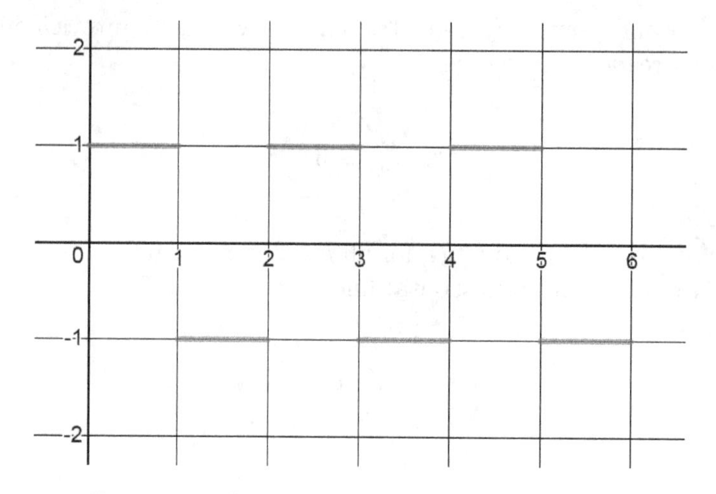

Figure 1.6: System of alternating horizontal lines.

In Figure 1.6, the first horizontal line $y = 1$ is on the interval $(0, 1)$ switches to the horizontal line $y = -1$ on the interval $(1, 2)$ and so on. This is described by the corresponding **alternating pattern**:

$$1, -1, 1, -1, 1, -1, \ldots. \tag{1.5}$$

Hence for all $n \geq 0$, you reformulate (1.5) as the cognate **piecewise function**:

$$y = \begin{cases} 1 & \text{if } x \in (0,1) \\ -1 & \text{if } x \in (1,2) \\ 1 & \text{if } x \in (2,3) \\ -1 & \text{if } x \in (3,4) \\ \vdots & \\ 1 & \text{if } x \in (2n, 2n+1) \\ -1 & \text{if } x \in (2n+1, 2n+2) \\ \vdots & \end{cases} \tag{1.6}$$

For all $n \geq 0$, you can alternatively reformulate (1.6) as the corresponding **alternating sequence** or an **alternating pattern**

$$y = (-1)^n \quad \text{if} \quad x \in (n, n+1).$$

Figure 1.7 traces an oscillatory system of blue diagonal lines with positive slope 1 and green diagonal lines with negative slope -1.

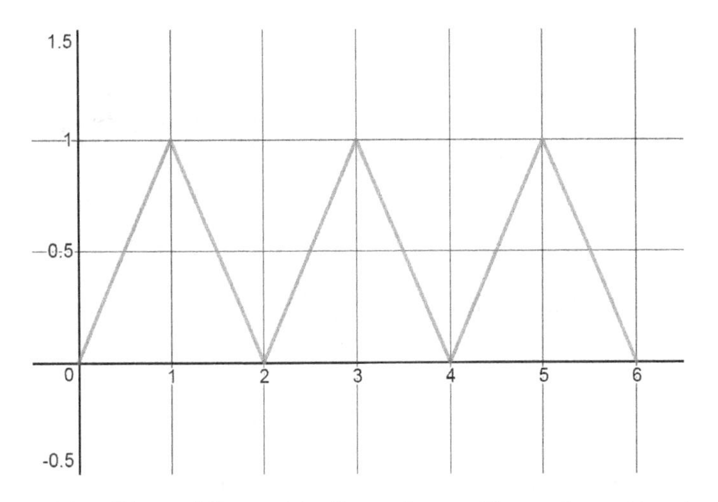

Figure 1.7: Diagonal lines with alternating positive and negative slopes.

Figure 1.7 commences with the diagonal line $y = x$ with slope 1 on the interval $[0, 1]$ and switches to the diagonal line $y = 2 - x$ with slope -1 on the interval $[1, 2]$, then switches to the diagonal line $y = x - 2$ with slope 1 on the interval $[2, 3]$, then switches to the diagonal green line $y = 4 - x$ with slope -1 on the interval $[3, 4]$ and so on.

Next, observe that the x-intercepts of the blue lines rise in increments of 2 and list all the consecutive positive even integers described by the following sequence:

$$0,\ 2,\ 4,\ 6,\ 8,\ \ldots,$$

while the x-intercepts of the green lines also rise in increments of 2 described by the corresponding sequence:

$$2,\ 4,\ 6,\ 8,\ 10,\ \ldots.$$

The left end-points of the restricted intervals of the blue lines are depicted by the following sequence:

$$0,\ 2,\ 4,\ 6,\ 8,\ \ldots. \tag{1.7}$$

The right end-points of the restricted intervals of the green lines are rendered by the associated sequence:

$$1,\ 3,\ 5,\ 7,\ 9,\ \ldots. \tag{1.8}$$

The left end-points of the restricted intervals of the green lines are depicted by the following sequence:

$$1,\ 3,\ 5,\ 7,\ 9,\ \ldots. \tag{1.9}$$

The right end-points of the restricted intervals of the green lines are rendered by the associated sequence:

$$2,\ 4,\ 6,\ 8,\ 10,\ \ldots. \tag{1.10}$$

Via (1.7), (1.8), (1.9) and (1.10), for all $n \geq 0$ you acquire:

$$y = \begin{cases} x - 0 & \text{if } x \in [0, 1] \\ 2 - x & \text{if } x \in [1, 2] \\ x - 2 & \text{if } x \in [2, 3] \\ 4 - x & \text{if } x \in [3, 4] \\ x - 4 & \text{if } x \in [4, 5] \\ 6 - x & \text{if } x \in [5, 6] \\ \vdots & \\ x - 2n & \text{if } x \in [2n, 2n+1] \\ (2n+2) - x & \text{if } x \in [2n+1, 2n+2] \\ \vdots & \end{cases} \tag{1.11}$$

Equation (1.11) is another example of an **alternating pattern** as two neighboring lines have slopes of opposite sign 1 and -1. Akin piecewise functions can be constructed and will be left as end of chapter exercises in Chapter 2.

Next, we will direct our focus to formulation of patterns (sequences) analytically as a list of specific values that you analogously encountered in (1.4).

1.3 Analytical Formulations

This section will focus on the examination and formulation of patterns by describing them analytically with a formula or with a set of formulas that outlines a precise list of assigned values. First, we will introduce several definitions.

Definition 1. For $n \in \mathbb{N}$, we define a **finite sequence** $\{x_i\}_{i=1}^{n}$ as a finite list of n values as:

$$\{x_i\}_{i=1}^{n} = x_1,\ x_2,\ x_3,\ x_4,\ \ldots,\ x_{n-1},\ x_n, \tag{1.12}$$

where 1 is the **starting index** and n is the **terminating index**. You can reformulate (1.12) by downshifting the starting and terminating indices by 1 less as:

$$\{x_i\}_{i=0}^{n-1} = x_0,\ x_1,\ x_2,\ x_3,\ \ldots,\ x_{n-2},\ x_{n-1}.$$

Similarly, by upshifting the starting and terminating indices by 1 more, reformulate (1.12) as:

$$\{x_i\}_{i=2}^{n+1} = x_2,\ x_3,\ x_4,\ x_5,\ \ldots,\ x_n,\ x_{n+1}.$$

Definition 2. We define an **infinite sequence** $\{x_n\}_{n=1}^{\infty}$ as a non-terminating list of values in the form:

$$\{x_n\}_{n=1}^{\infty} = x_1,\ x_2,\ x_3,\ x_4,\ \ldots,\ x_n,\ \ldots. \tag{1.13}$$

The aim is to detect the contrasting patterns such as: linear sequences, geometric sequences, summation-type sequences, product-type sequences, factorial, alternating sequence and piecewise sequences later in this chapter and in thorough attributes in Chapters 2 and 3. The succeeding **linear sequence** or a linear pattern lists all the consecutive positive integers (natural numbers $\mathbb{N}$) starting with 1 as we detected in (1.4):

$$\{n\}_{n=1}^{\infty} = 1,\ 2,\ 3,\ 4,\ 5,\ 6,\ 7,\ 8,\ \ldots. \tag{1.14}$$

Observe that (1.14) is convened by adding a 1 from neighbor to neighbor. In fact, the difference between two neighbors in (1.14) is always 1. (1.14) is a special case of a **linear pattern** when the difference between two neighbors is always a constant. For instance, you can formulate analogous linear patterns such as the consecutive positive even integers starting with 2 or the positive multiples of 2:

$$\begin{aligned}\{2n\}_{n=1}^{\infty} &= 2,\ 4,\ 6,\ 8,\ 10,\ 12,\ 14,\ \ldots \\ &= 2\cdot 1,\ 2\cdot 2,\ 2\cdot 3,\ 2\cdot 4,\ 2\cdot 5,\ \ldots.\end{aligned} \tag{1.15}$$

Note, that the difference between two neighbors in (1.15) is always 2 and (1.15) lists the consecutive positive even integers. Akin to (1.15), the consequent sequence enumerates the consecutive positive odd integers starting with 1:

$$\begin{aligned}\{2n-1\}_{n=1}^{\infty} &= 1,\ 3,\ 5,\ 7,\ 9,\ 11,\ 13,\ \ldots \\ &= 2\cdot 1\ -\ 1,\ 2\cdot 2\ -\ 1,\ 2\cdot 3\ -\ 1,\ 2\cdot 4\ -\ 1,\ \ldots\ .\end{aligned} \tag{1.16}$$

The next example examines a linear sequence that recites multiples of 3.

Example 1.1. Write a formula of the following sequence:

$$3,\ 6,\ 9,\ 12,\ 15,\ 18,\ 21,\ \ldots. \tag{1.17}$$

Solution: First note that (1.17) starts with 3 and shifts from neighbor to neighbor by adding a 3. You can then reformulate (1.17) as:

$$3\cdot 1,\ 3\cdot 2,\ 3\cdot 3,\ 3\cdot 4,\ 3\cdot 5,\ 3\cdot 6,\ 3\cdot 7,\ \ldots. \tag{1.18}$$

Hence, via (1.18) you obtain the associated formula:

$$\{3n\}_{n=1}^{\infty}. \tag{1.19}$$

By downshifting the starting index of (1.19) by 1 you obtain:

$$\{3n\}_{n=1}^{\infty} = \{3(n+1)\}_{n=0}^{\infty} = \{3n+3\}_{n=0}^{\infty}.$$

The upcoming example analyzes a Linear Sequence that enumerates 1 less than the multiples of 4.

Example 1.2. Write a formula of the following sequence:

$$3,\ 7,\ 11,\ 15,\ 19,\ 23,\ 27,\ \ldots. \tag{1.20}$$

Solution: First, note that (1.20) starts with 3 and shifts from neighbor to neighbor by adding a 4. Next, observe that (1.20) lists values that are one less than multiples of 4. You can then reformulate (1.20) as:

$$4\cdot 1-1,\ 4\cdot 2-1,\ 4\cdot 3-1,\ 4\cdot 4-1,\ 4\cdot 5-1,\ 4\cdot 6-1,\ 4\cdot 7-1,\ \ldots. \quad (1.21)$$

Hence, via (1.21) you obtain the associated formula:

$$\{4n-1\}_{n=1}^{\infty}. \quad (1.22)$$

By downshifting the starting index of (1.22) by 1 you obtain:

$$\{4n-1\}_{n=1}^{\infty} = \{4(n+1)-1\}_{n=0}^{\infty} = \{4n+3\}_{n=0}^{\infty}.$$

Analogous to (1.14), (1.15), (1.16), (1.17) and (1.20), you can formulate additional linear patterns. Supplemental examples of linear sequences will be analyzed in Chapter 3. Next, we will shift our focus on geometric sequences.

Definition 3. For $n \geq 0$, we define a **finite geometric sequence** consisting of $n+1$ values as:

$$\{a\cdot r^i\}_{i=0}^{n} = a\cdot r^0,\ a\cdot r^1,\ a\cdot r^2,\ a\cdot r^3,\ a\cdot r^4,\ a\cdot r^5,\ \ldots,\ a\cdot r^n, \quad (1.23)$$

where a is the **starting term** of (1.23) and r is the **multiplicative factor** in (1.23). Note that the pattern in (1.23) is generated by multiplying by r from neighbor to neighbor. The quotient of two neighbors in (1.23) is always r.

Definition 4. For $n \geq 0$, we define an **infinite geometric sequence** as:

$$\{a\cdot r^n\}_{n=0}^{\infty} = a\cdot r^0,\ a\cdot r^1,\ a\cdot r^2,\ a\cdot r^3,\ a\cdot r^4,\ a\cdot r^5,\ \ldots, \quad (1.24)$$

where a is the **starting term** and r is the **multiplicative factor**.

The upcoming example is a geometric sequence that recites the powers of 3.

Example 1.3. Write a formula of the following sequence:

$$3,\ 9,\ 27,\ 81,\ 243,\ 729,\ 2{,}187,\ \ldots. \tag{1.25}$$

Solution: First, (1.25) starts with 3 and shifts from neighbor to neighbor by multiplying by 3. Next, you reformulate (1.25) as:

$$3^1,\ 3^2,\ 3^3,\ 3^4,\ 3^5,\ 3^6,\ 3^7,\ \ldots. \tag{1.26}$$

Hence, via (1.26) you obtain the associated formula:

$$\{3^n\}_{n=1}^{\infty}. \tag{1.27}$$

By downshifting the starting index of (1.27) by 1 you obtain:

$$\{3^n\}_{n=1}^{\infty} = \{3^{n+1}\}_{n=0}^{\infty}.$$

The consequent example is geometric sequence that recites the powers of $\sqrt{2}$.

Example 1.4. Write a formula of the following sequence:

$$2,\ \sqrt{8},\ 4,\ \sqrt{32},\ 8,\ \sqrt{128},\ 16,\ \ldots. \tag{1.28}$$

Solution: First, (1.28) starts with 2 and shifts from neighbor to neighbor by multiplying by $\sqrt{2}$. Next, reformulate (1.28) as:

$$\left[\sqrt{2}\right]^2,\ \left[\sqrt{2}\right]^3,\ \left[\sqrt{2}\right]^4,\ \left[\sqrt{2}\right]^5,\ \left[\sqrt{2}\right]^6,\ \left[\sqrt{2}\right]^7,\ \left[\sqrt{2}\right]^8,\ \ldots. \tag{1.29}$$

Hence, via (1.29) you obtain the associated formula:

$$\left\{\left[\sqrt{2}\right]^n\right\}_{n=2}^{\infty}. \tag{1.30}$$

By downshifting the starting index of (1.30) by 1 you obtain:

$$\left\{\left[\sqrt{2}\right]^n\right\}_{n=2}^{\infty} = \left\{\left[\sqrt{2}\right]^{n+1}\right\}_{n=1}^{\infty}.$$

Figures 1.3 and 1.4 are described by a geometric sequence. In Figure 1.3, the green triangle's dimensions are half of the principal blue triangle's dimensions and so on. The succeeding Section 1.3.1 will focus on applications of a geometric sequence in paper folding.

1.3.1 *Geometric Sequences and Paper Folding*

This section will study applications of a geometric sequence in paper folding. Figure 1.8 starts with the largest square with area 1. The square is folded in half vertically first, then horizontally, etc. This rectangular folding is done 8 times as illustrated in Figure 1.8.

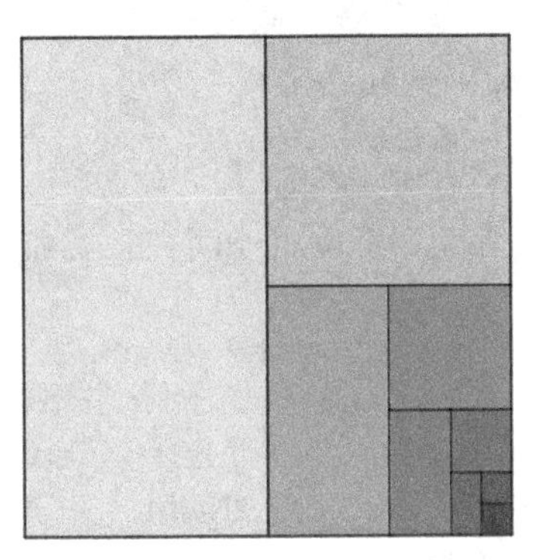

Figure 1.8: Square paper folding vertically and horizontally.

In Figure 1.8, the largest square is folded in half vertically first, then horizontally, etc. This rectangular folding is performed 8 times. You first cut the main square in half vertically, then horizontally, then vertically, then horizontally, etc. Diminishing rectangles and their associated areas are shaded with darker shades of blue. During each fold, you reduce the area by half and hence generate the following geometric sequence:

$$1, \frac{1}{2}, \left(\frac{1}{2}\right)^2, \left(\frac{1}{2}\right)^3, \left(\frac{1}{2}\right)^4, \left(\frac{1}{2}\right)^5, \left(\frac{1}{2}\right)^6, \left(\frac{1}{2}\right)^7, \left(\frac{1}{2}\right)^8$$
$$= \left\{\left(\frac{1}{2}\right)^i\right\}_{i=0}^{8}.$$

In Chapter 2, you will sketch the fractal shaped presented in Figure 1.8 with a system of horizontal and vertical lines on the corresponding restricted intervals.

Figure 1.9 examines triangular paper folding in diagonal and vertical directions and presents the largest right triangle with area 1.

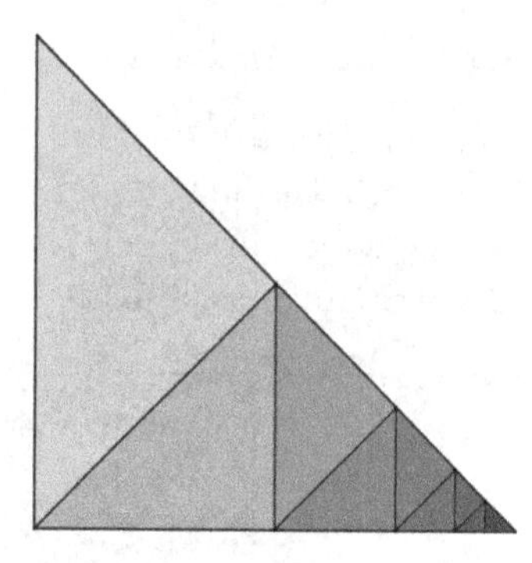

Figure 1.9: Triangular paper folding diagonally and vertically.

In Figure 1.9, you first slice the principle right triangle in half diagonally, then vertically, etc. Diminishing right triangles emerge with the darker shades of blue. This triangular folding is done 8 times and each time you fold you shrink the area by half and thus produce the following geometric sequence:

$$1,\ \frac{1}{2},\ \left(\frac{1}{2}\right)^2,\ \left(\frac{1}{2}\right)^3,\ \left(\frac{1}{2}\right)^4,\ \left(\frac{1}{2}\right)^5,\ \left(\frac{1}{2}\right)^6,\ \left(\frac{1}{2}\right)^7,\ \left(\frac{1}{2}\right)^8$$
$$= \left\{\left(\frac{1}{2}\right)^i\right\}_{i=0}^{8}.$$

Next, we will direct our focus on recursive sequences.

1.4 Recursive Sequences

This section's aims are to formulate a specific sequence recursively as an **initial value problem**, where the initial value is the starting value of a sequence. Analogous to Examples (1.1) and (1.2), the upcoming example examines a linear sequence by shifting from neighbor to neighbor by adding a 2.

Example 1.5. Write a recursive formula for:

$$1,\ 3,\ 5,\ 7,\ 9,\ 11,\ 13,\ \ldots. \tag{1.31}$$

Solution: Observe (1.31) starts with 1 and transitions from neighbor to neighbor by adding a 2. Recursively and inductively you formulate

(1.31) as:

$$\begin{aligned} x_0 &= 1, \\ x_1 &= x_0 + 2 = 1 + 2 = 3, \\ x_2 &= x_1 + 2 = 3 + 2 = 5, \\ x_3 &= x_2 + 2 = 5 + 2 = 7, \\ x_4 &= x_3 + 2 = 7 + 2 = 9, \\ &\vdots \end{aligned} \tag{1.32}$$

Via (1.32), for all $n \geq 0$ you obtain the following **initial value problem**:

$$\begin{cases} x_{n+1} = x_n + 2, \\ x_0 = 1. \end{cases}$$

Example (1.5) leads you to the examination of summation-type sequences.

1.4.1 *Summation-type Sequences*

Definition 5. For $n \in \mathbb{N}$, we define the **Sum** consisting of n values with the symbol $\sum$ (sigma notation) as:

$$S = x_1 + x_2 + x_3 + \cdots + x_n = \sum_{i=1}^{n} x_i. \tag{1.33}$$

The left-hand side of (1.33) is the **expanded form** of S and the right-hand side of (1.33) is the **factored form** of S expressed in the sigma notation.

The next example adds consecutive positive integers starting with 1.

Example 1.6. Write a recursive formula for:

$$1,\ 3,\ 6,\ 10,\ 15,\ 21,\ 28,\ \ldots. \tag{1.34}$$

Solution: (1.34) starts at 1 and you shift to the next term by adding a 2, then add a 3 to the consecutive neighbor and so on. Recursively and inductively formulate (1.34) as:

$$\begin{aligned} x_0 &= 1, \\ x_1 &= x_0 + 2 = x_0 + (0+2) = 1 + 2 = 3, \\ x_2 &= x_1 + 3 = x_1 + (1+2) = 3 + 3 = 6, \\ x_3 &= x_2 + 4 = x_2 + (2+2) = 6 + 4 = 10, \\ x_4 &= x_3 + 5 = x_3 + (3+2) = 10 + 5 = 15, \\ x_5 &= x_4 + 6 = x_4 + (4+2) = 15 + 6 = 21, \\ &\vdots \end{aligned} \tag{1.35}$$

Via (1.35), for all $n \geq 0$ you obtain the following **initial value problem:**

$$\begin{cases} x_{n+1} = x_n \ + \ (n+2), \\ \quad x_0 = 1. \end{cases} \tag{1.36}$$

Figure 1.10 resembles 1.36 geometrically as the following system of step-shaped squares.

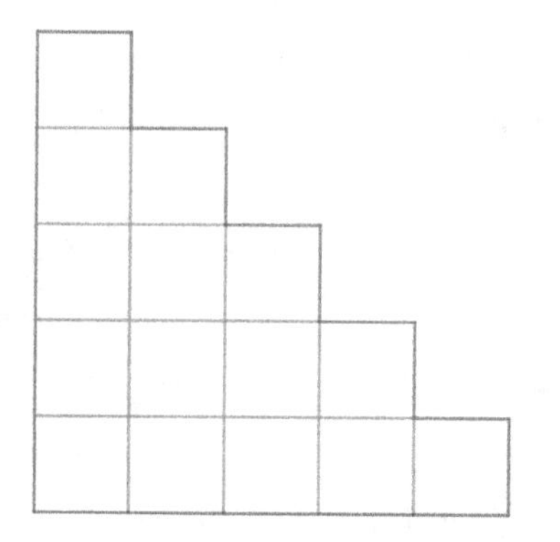

Figure 1.10: A system of step-shaped squares.

In Figure 1.10, the first row has one square, the second row has 2 squares, the third row has 3 squares and so on. Hence, you formulate (1.36) as the following summation of consecutive positive integers starting with 1 as:

$$1 \ + \ 2 \ + \ 3 \ + \ 4 \ + \ 5 \ + \ 6 \ + \cdots + n = \sum_{i=1}^{n} i. \tag{1.37}$$

The next Section 1.4.2 will examine the Fibonacci sequence.

1.4.2 *The Fibonacci Sequence*

This section's aims are to examine the special case of recursive sequences as the structure of the following **Fibonacci numbers** or the **Fibonacci sequence**:

$$1,\ 1,\ 2,\ 3,\ 5,\ 8,\ 13,\ 21,\ \ldots. \tag{1.38}$$

In (1.38), you commence by adding the first two consecutive terms 1 and 1. You then proceed by adding the next two consecutive terms 1 and 2, and so on. Therefore, formulate (1.38) as the corresponding recursive relation:

$$\begin{aligned}
&x_0 = 1,\\
&x_1 = 1,\\
&x_2 = x_0 + x_1 = 1 + 1 = 2,\\
&x_3 = x_1 + x_2 = 1 + 2 = 3,\\
&x_4 = x_2 + x_3 = 2 + 3 = 5,\\
&x_5 = x_3 + x_4 = 3 + 5 = 8,\\
&\vdots
\end{aligned}$$

The **Fibonacci sequence** emerges in nature's patterns such as the structure of pine cones, trees, leaves, sunflowers, sea shells, storm systems and the galaxies' formations. They often appear in the corresponding spiraling shape illustrated in the following sketch:

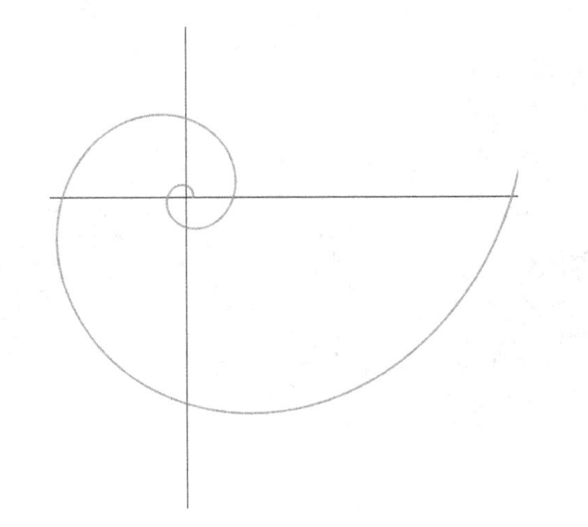

Analogous to the spiraling shape in the sketch above, the consequent photograph describes the spiral Fibonacci pattern of the sea shell's structure:

The following photograph captures the spiral Fibonacci pattern of the storm cloud's formation:

The next Section 1.4.3 will focus on product-type sequences.

1.4.3 *Product-Type Sequences*

This section's aims are to investigate Product-Type Sequences. Akin to Example (1.3), the consequent example describes a Geometric Sequence recursively by multiplying from neighbor to neighbor.

Example 1.7. Write a recursive formula for:

$$5,\ 10,\ 20,\ 40,\ 80,\ 160,\ 320,\ \dots. \tag{1.39}$$

Solution: Observe that (1.39) starts with 5 and transitions from neighbor to neighbor by multiplying by 2. Thus, recursive and inductively formulate (1.31) as:

$$\begin{aligned}
x_0 &= 5,\\
x_1 &= 2 \cdot x_0 = 2 \cdot 5 = 10,\\
x_2 &= 2 \cdot x_1 = 2 \cdot 10 = 20,\\
x_3 &= 2 \cdot x_2 = 2 \cdot 20 = 40,\\
x_4 &= 2 \cdot x_3 = 2 \cdot 40 = 80,\\
x_5 &= 2 \cdot x_4 = 2 \cdot 80 = 160,\\
x_6 &= 2 \cdot x_5 = 2 \cdot 160 = 320,\\
&\vdots
\end{aligned} \tag{1.40}$$

Via (1.40), for all $n \geq 0$ you obtain the following **initial value problem**:

$$\begin{cases} x_{n+1} = 2 \cdot x_n, \\ \quad x_0 = 5. \end{cases}$$

Next, let's transition to products with the corresponding definition.

Definition 6. For $n \in \mathbb{N}$, we define the **product** of n values with the symbol $\prod$ as:

$$P = x_1 \cdot x_2 \cdot x_3 \cdot \ \cdots \ \cdot x_n = \prod_{i=1}^{n} x_i. \tag{1.41}$$

The left-hand side of (1.41) is the **expanded form** of P and the right-hand side of (1.41) is the **factored form** of P expressed in the product notation. The upcoming example multiplies powers of 2.

Example 1.8. Write a formula of the following product-type sequence:

$$2^1 \cdot 2^3 \cdot 2^5 \cdot 2^7 \cdot \cdots \cdot 2^{2n-3} \cdot 2^{2n-1}. \tag{1.42}$$

Solution: The starting index of (1.42) is 1 and the terminating index of (1.42) is n. Then reformulate (1.42) in the factored form as:

$$2^1 \cdot 2^3 \cdot 2^5 \cdot 2^7 \cdot 2^9 \cdot \cdots \cdot 2^{2n-3} \cdot 2^{2n-1} = \prod_{i=1}^{n} 2^{2i-1}.$$

The upcoming example presents the **factorial pattern** as a special case of the product-type sequences.

Example 1.9. The **factorial pattern** is defined as a product of consecutive positive integers starting with 1 as:

$$\begin{aligned}
&0! = 1,\\
&1! = 1,\\
&2! = 2 \cdot 1 = 2 \cdot 1!,\\
&3! = 3 \cdot 2 \cdot 1 = 3 \cdot 2!,\\
&4! = 4 \cdot 3 \cdot 2 \cdot 1 = 4 \cdot 3!,\\
&\vdots\\
&n! = n \cdot (n-1) \cdot (n-2) \cdot \cdots \cdot 2 \cdot 1 = n \cdot (n-1)! = \prod_{i=1}^{n} i.
\end{aligned} \tag{1.43}$$

The geometric sequence becomes a special case of the factorial when you multiply by the same constant r. You will formulate the factorial as a recursive sequence in Chapter 5. The factorial also naturally emerges in the construction of each row of the **Pascal's triangle** illustrated in Figure 1.11.

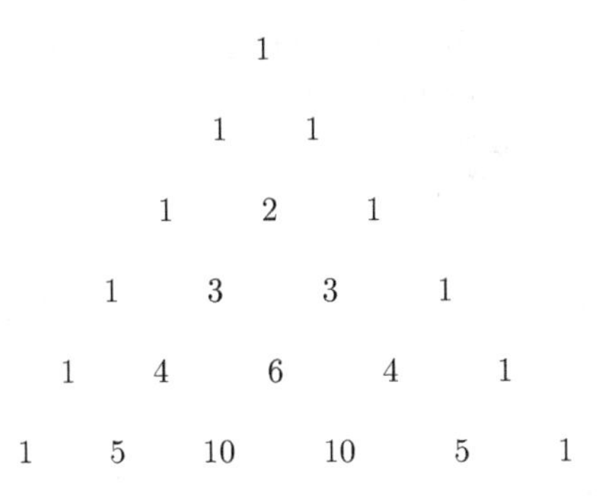

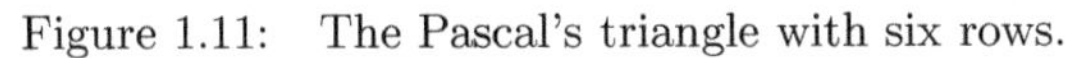
Figure 1.11: The Pascal's triangle with six rows.

In Chapter 5, you will examine deeper features of the **Pascal's triangle** in Figure 1.11 by deciphering the triangle's rows and diagonals. You will apply identities expressed in combinations while formulating and proving specific patterns of the **Pascal's triangle**.

The Factorial pattern in (1.43) is a special case of a piecewise sequence that can be reformulated as follows:

$$n! = \begin{cases} 1 & \text{if } n = 0, \\ \prod_{i=1}^{n} i & \text{if } n \in \mathbb{N}. \end{cases} \tag{1.44}$$

Equation (1.43) and (1.44) now guide you to piecewise sequences.

1.5 Piece-wise Sequences

This section's objectives are to examine piecewise sequences that consist of two or more sub-sequences.

Definition 7. For all $n \geq 0$, we define a **piecewise sequence** $\{x_n\}_{n=0}^{\infty}$ that consists of two sub-sequences $\{a_n\}_{n=0}^{\infty}$ and $\{b_n\}_{n=1}^{\infty}$ as:

$$a_0, \; b_1, \; a_2, \; b_3, \; \ldots. \tag{1.45}$$

Note that (1.45) can be expressed as:

$$\{x_n\}_{n=0}^{\infty} = \begin{cases} a_n & \text{if } n = 0, 2, 4, 6, \ldots, \\ b_n & \text{if } n = 1, 3, 5, 7, \ldots. \end{cases} \tag{1.46}$$

Equation (1.46) can be either decoupled into even-indexed and odd-indexed sub-sequences, positive and negative sub-sequences, and into alternating and non-alternating sub-sequences. The first example decomposes an alternating sequence into positive and negative sub-sequences.

Example 1.10. For all $n \in \mathbb{N}$, reformulate the alternating sequence:

$$\{(-1)^{n+1}n\}_{n=1}^{\infty} = 1,\ -2,\ 3,\ -4,\ 5,\ -6,\ 7,\ -8,\ \ldots. \tag{1.47}$$

as a piecewise sequence.

Solution: You group (1.47) into blue even-indexed terms and into green odd-indexed terms as:

$$1,\ -2,\ 3,\ -4,\ 5,\ -6,\ \ldots. \tag{1.48}$$

Via (1.48), for all $n \geq 0$, reformulate (1.48) as:

$$\{x_n\}_{n=0}^{\infty} = \begin{cases} [n+1] & \text{if } n = 0, 2, 4, 6, \ldots, \\ -[n+1] & \text{if } n = 1, 3, 5, 7, \ldots. \end{cases}$$

The upcoming example decouples the given sequence into even-indexed and odd-indexed linear sub-sequences.

Example 1.11. Write a piecewise formula of the following sequence:

$$0,\ 1,\ 4,\ 5,\ 8,\ 9,\ 12,\ 13,\ \ldots. \tag{1.49}$$

Solution: First break up (1.49) into two blue and green linear sub-sequences as follows:

$$0,\ 1,\ 4,\ 5,\ 8,\ 9,\ 12,\ 13,\ \ldots. \tag{1.50}$$

Via (1.50), for $n \geq 0$ you acquire:

$$\{x_n\}_{n=0}^{\infty} = \begin{cases} 2n & \text{if } \ n = 0, 2, 4, 6, \ldots, \\ 2n-1 & \text{if } \ n = 1, 3, 5, 7, \ldots. \end{cases}$$

The succeeding example decomposes the assigned sequence into even-indexed and odd-indexed geometric sub-sequences.

Example 1.12. Write a piecewise formula of the following sequence:

$$1,\ C_0,\ C_0C_1,\ C_0^2C_1,\ C_0^2C_1^2,\ C_0^3C_1^2,\ C_0^3C_1^3,\ C_0^4C_1^3,\ \ldots. \tag{1.51}$$

Solution: First decompose (1.51) into two blue and green geometric sub-sequences as:

$$1,\ C_0,\ [C_0C_1],\ C_0[C_0C_1],\ [C_0C_1]^2,\ C_0[C_0C_1]^2,\ [C_0C_1]^3,\ C_0[C_0C_1]^3,\ \ldots. \tag{1.52}$$

Via (1.52), for $n \geq 0$ you obtain:

$$\{x_n\}_{n=0}^{\infty} = \begin{cases} [C_0C_1]^{\frac{n}{2}} & \text{if } n = 0, 2, 4, 6, \ldots, \\ C_0[C_0C_1]^{\frac{n-1}{2}} & \text{if } n = 1, 3, 5, 7, \ldots. \end{cases}$$

You will seek and examine supplemental examples of piecewise sequences and their applications in Chapters 2 and 3. The upcoming section will shift your focus on periodic sequences or patterns.

1.6 Periodic Cycles

This section's aims are to examine the characteristics of periodic sequences or patterns. Periodic sequences are special cases of piecewise sequences that consist of two or more fragments as you encountered in Figures 1.6 and 1.7 and describe periodic behavior. In some instances, you will encounter alternating patterns. Let's commence with the following definition.

Definition 8. The sequence $\{x_n\}_{n=0}^{\infty}$ is periodic with **period-*p*,** $(p \geq 2)$, provided that for all $n \geq 0$:

$$x_{n+p} = x_n. \tag{1.53}$$

From Figure 1.6, you acquired the following alternating and piecewise period-2 pattern:

$$1,\ -1,\ 1,\ -1,\ 1,\ -1,\ \ldots. \tag{1.54}$$

For all $n \geq 0$, formulate (1.54) as the following piecewise formula:

$$(-1)^n = \begin{cases} 1 & \text{if } n = 0, 2, 4, 6, \ldots, \\ -1 & \text{if } n = 1, 3, 5, 7, \ldots. \end{cases}$$

The graph in Figure 1.12, depicts the assigned alternating period-2 cycle in (1.54).

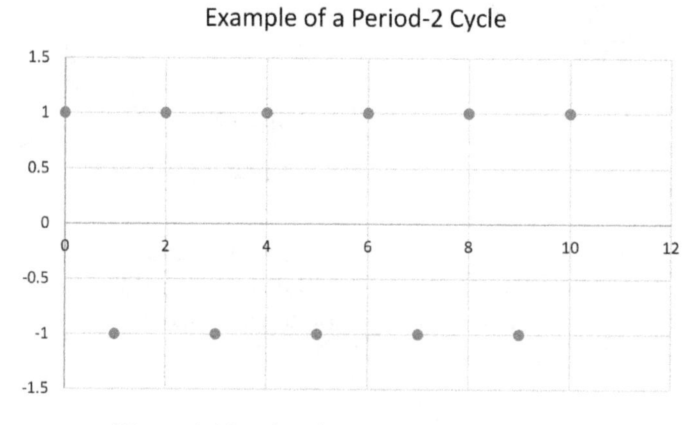

Figure 1.12: An alternating period-2 cycle.

The corresponding graph in Figure 1.13.

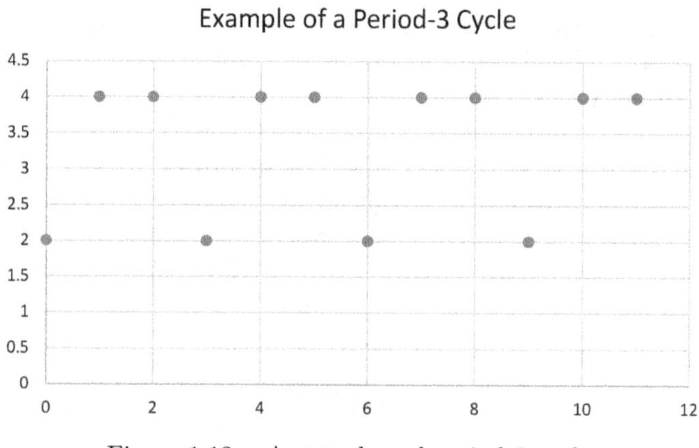

Figure 1.13: A step-shaped period-3 cycle.

Sketches the associated ascending step-shaped period-3 cycle:

$$2,\ 4,\ 4,\ 2,\ 4,\ 4,\ \ldots,$$

which you can also describe as the following piecewise sequence:

$$\{x_n\}_{n=0}^{\infty} = \begin{cases} 2 & \text{if } n = 3k, \\ 4 & \text{if } n = 3k+1, \\ 4 & \text{if } n = 3k+2. \end{cases}$$

Figure 1.13 renders a special case of periodic attributes of Eq. (7.1) in Chapter 7 when $x_0 = 1$, $b_0 = 2$, $b_1 = 2$ and $b_2 = -4$. You will seek additional examples of periodicity in Chapter 7 and compare the similarities and differences with periodicity in Figure 1.13.

The graph in Figure 1.14.

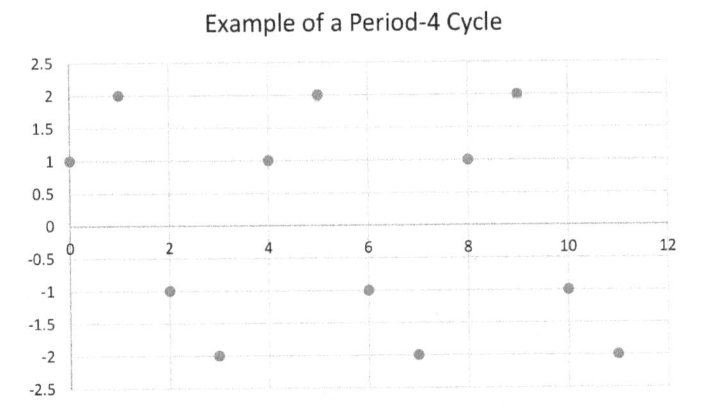

Figure 1.14: An alternating period-4 cycle.

describes the corresponding alternating period-4 cycle:

$$1,\ 2,\ -1,\ -2,\ 1,\ 2,\ -1,\ -2,\ \ldots,$$

which you can express as the analogous piecewise pattern:

$$\{x_n\}_{n=0}^{\infty} = \begin{cases} 1 & \text{if } n = 4k, \\ 2 & \text{if } n = 4k+1, \\ -1 & \text{if } n = 4k+2, \\ -2 & \text{if } n = 4k+3. \end{cases}$$

Figure 1.14 is a special case of Eq. (7.1) in Chapter 7 when $x_0 = 1$, $a_0 = 2$ and $a_1 = -0.5$. You will examine supplemental examples of analogous periodic traits in Chapter 7 that mimic the pattern you see in Figure 1.14. Figures 1.12 and 1.14 resemble alternating patterns.

Definition 9. The sequence $\{x_n\}_{n=0}^{\infty}$ is an **alternating periodic sequence** with period-$2k$ ($k \geq 1$), if $x_{n+k} = -x_n$ for all $n \geq 0$.

The sketch in Figure 1.15 traces an alternating period-8 cycle.

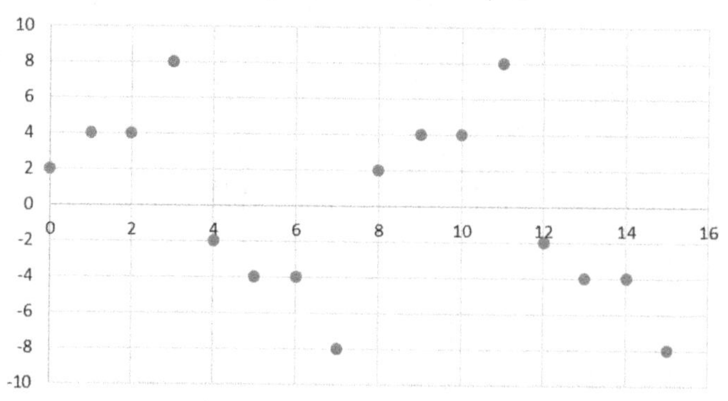

Figure 1.15: An alternating period-8 cycle.

Note, that Figure 1.12 describes an alternating period-2 cycle as $x_{n+1} = -x_n$ for all $n \geq 0$. Next, observe that Figure 1.14 presents an alternating period-4 cycle as $x_{n+2} = -x_n$ for all $n \geq 0$. Figure 1.15 portrays an alternating period-8 cycle as $x_{n+4} = -x_n$ for all $n \geq 0$. On the other hand, Figure 1.13 does not render an alternating period-3 cycle as all the terms of the cycle are positive. This will guide you to study further characteristics of alternating-periodic patterns formulated and characterized by recursive sequences in Chapter 7. The upcoming Section 1.6.1 will analyze supplemental periodic patterns with various shapes.

1.6.1 *Shapes of Periodic Cycles*

This section's aims are to portray periodic patterns with various shapes such as triangular-shaped and step-shaped. The first sketch in Figure 1.16 presents an **ascending triangular-shaped** period-3 cycle.

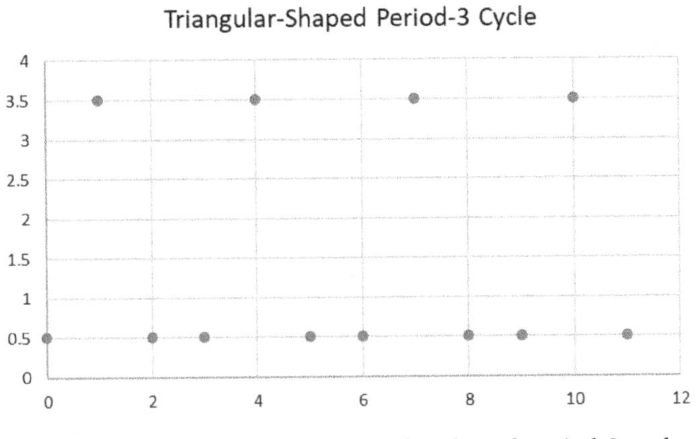

Figure 1.16: An ascending triangular-shaped period-3 cycle.

The consequent diagram in Figure 1.17 traces a **trapezoidal-shaped** period-4 cycle.

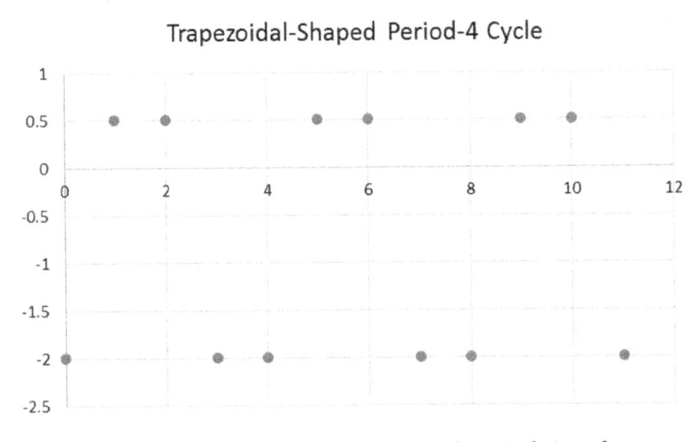

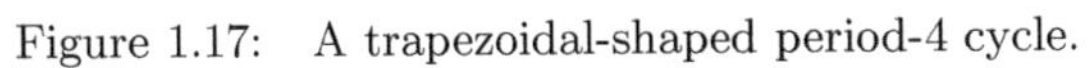

Figure 1.17: A trapezoidal-shaped period-4 cycle.

The next graph in Figure 1.18 describes an **ascending step-shaped** period-4 cycle.

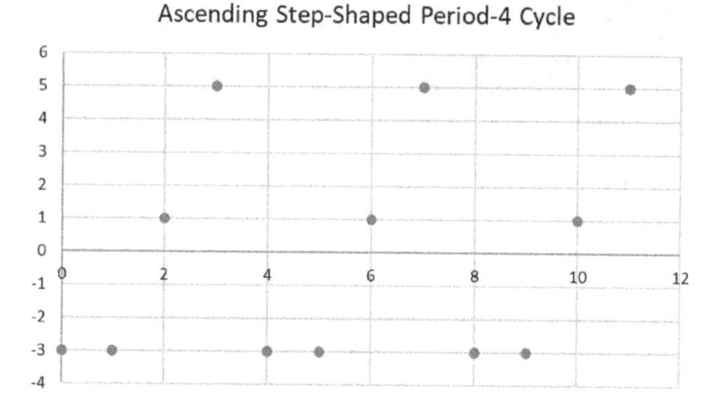

Figure 1.18: An ascending step-shaped period-4 cycle.

The sketch in Figure 1.19 portrays a **descending step-shaped** period-5 cycle.

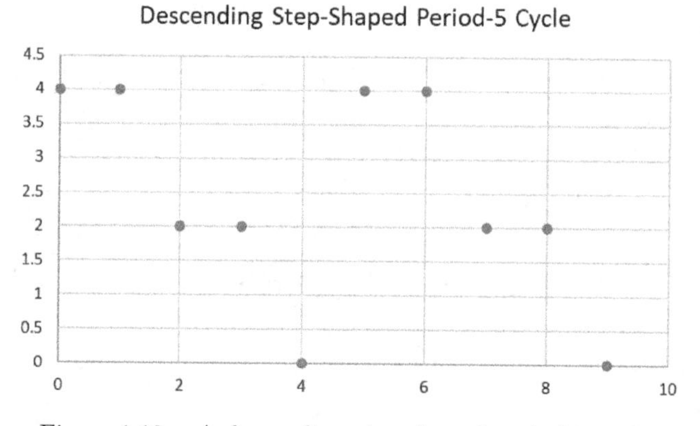

Figure 1.19: A descending step-shaped period-5 cycle.

1.7 Exercises

In problems 1–10, write a **formula** of each linear sequence:

1. 6, 12, 18, 24, 30, 36, 42, ...
2. 7, 14, 21, 28, 35, 42, 49, ...

3. $4, 12, 20, 28, 36, 44, 52, \ldots.$
4. $8, 10, 12, 14, 16, 18, 20, \ldots.$
5. $30, 35, 40, 45, 50, 55, 60, \ldots.$
6. $5, 7, 9, 11, 13, 15, 17, \ldots.$
7. $4, 7, 10, 13, 16, 19, 22, \ldots.$
8. $2, 11, 20, 29, 38, 47, 56, \ldots.$
9. $4, 11, 18, 25, 32, 39, 46, \ldots.$
10. $1, 5, 9, 13, 17, 21, 25, \ldots.$

In problems 11−18, write a **formula** of each geometric sequence:

11. $8, 16, 32, 64, 128, 256, 512, \ldots.$
12. $6, 12, 24, 48, 96, 192, 384, \ldots.$
13. $\frac{1}{9}, \frac{1}{3}, 1, 3, 9, 27, 81, \ldots.$
14. $\frac{3}{8}, \frac{3}{2}, 6, 24, 96, 384, 1{,}536, \ldots.$
15. $5, 15, 45, 135, 405, 1{,}215, 3{,}645, \ldots.$
16. $32, 8, 2, \frac{1}{2}, \frac{1}{8}, \frac{1}{32}, \ldots.$
17. $64, 48, 36, 27, \frac{81}{4}, \frac{343}{16}, \ldots.$
18. $9, 6, 4, \frac{8}{3}, \frac{16}{9}, \frac{32}{27}, \ldots.$

In problems 19−26, write a **formula** of each piecewise sequence:

19. $2, -4, 2, -4, 2, -4, \ldots.$
20. $-1, 2, 1, -2, -1, 2, 1, -2, \ldots.$
21. $3, 1, -1, 3, 1, -1, 3, 1, -1, \ldots.$
22. $2, 4, -1, -3, 2, 4, -1, -3, \ldots.$
23. $1, -4, 5, -8, 9, -12, 13, -16, \ldots.$
24. $-2, 4, -6, 8, -10, 12, -14, 16, \ldots.$
25. $1, -2, 4, -8, 16, -32, 64, \ldots.$
26. $2, 4, 6, -8, 10, 12, 14, -16, \ldots.$

In problems 27–30:

27. Write a **formula** of the piecewise function below:

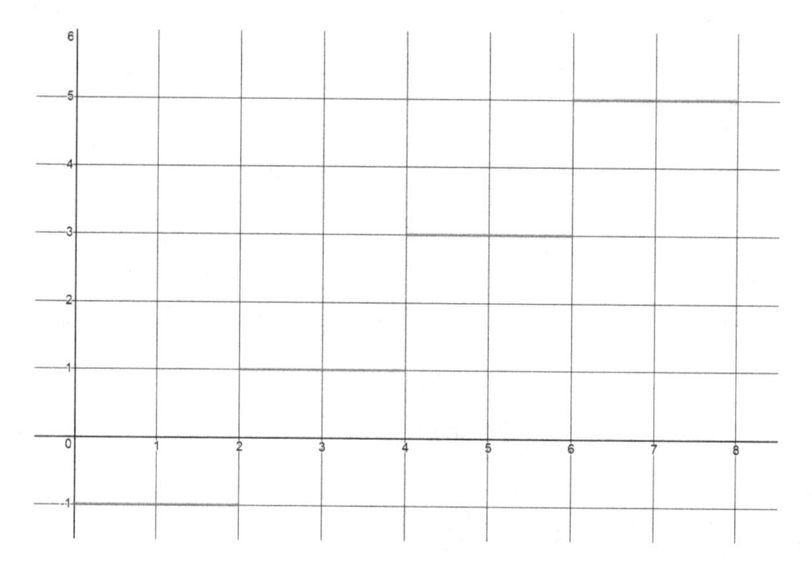

28. Write a **formula** of the piecewise function below:

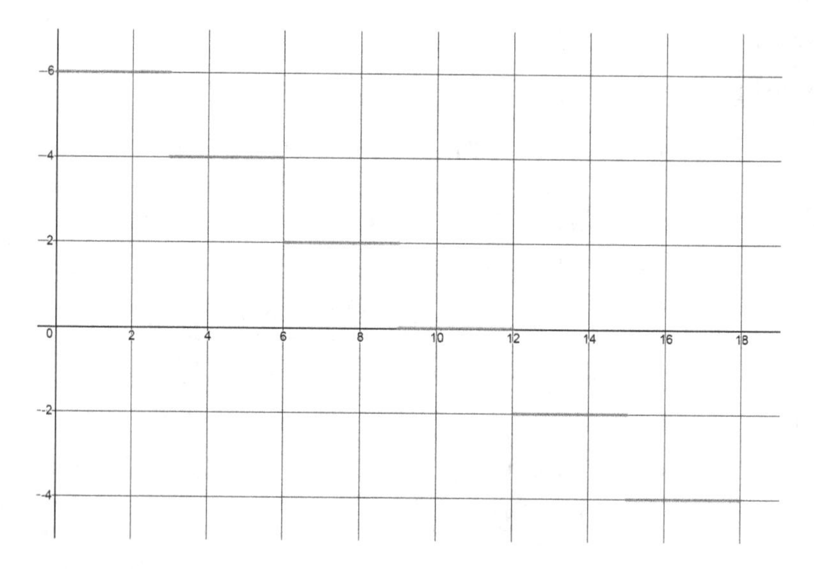

29. Write a **formula** of the piecewise function below:

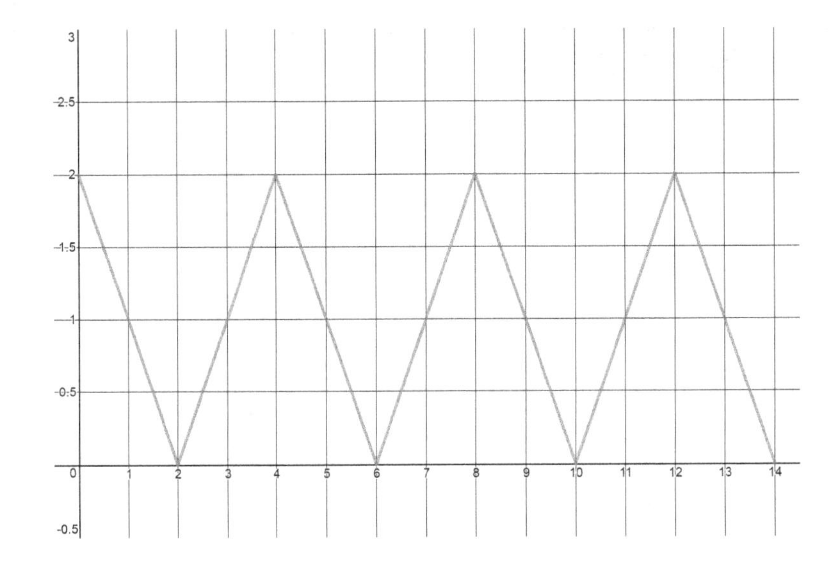

30. Write a **formula** of the piecewise function below:

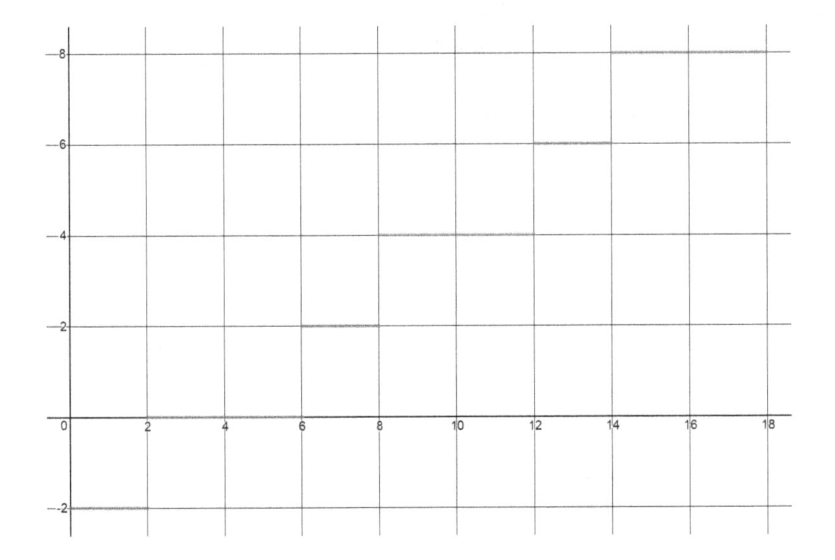

In problems 31–32:

31. Write a system of functions on restricted intervals that describe the right triangle below:

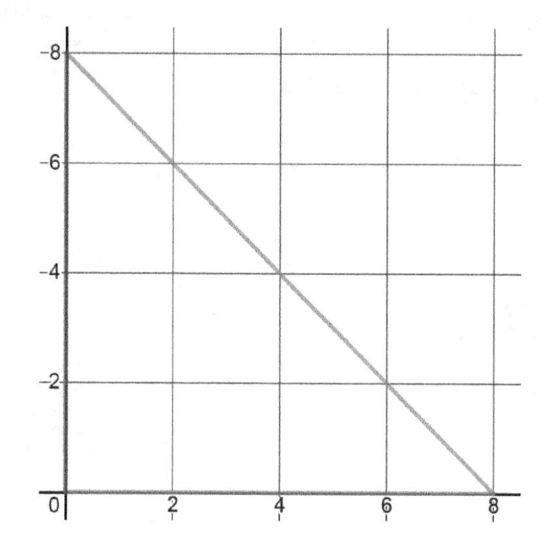

32. Write a system of functions on restricted intervals that describe the equilateral triangle below:

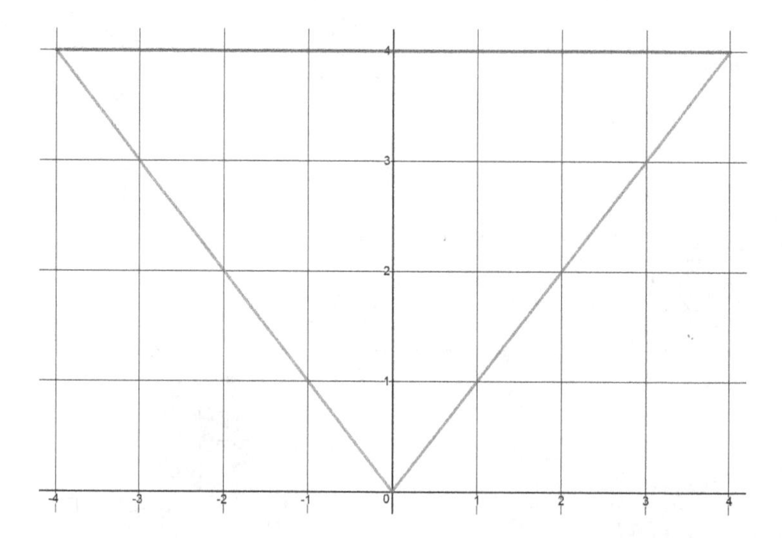

Chapter 2

Geometrical Configurations

This chapter's aims are to investigate numerous examples that describe how particular patterns (sequences) ensue from assorted geometrical arrangements of figures. You will examine configurations that repeat at the same scale, replicate at different scales (fractals), and with alternating patterns. During your exploration, you will ask the consequent questions: How do you distinguish the similarities and differences between various geometrical arrangements? How and why do certain arrangements guide you to formulations of specific patterns? You will also examine the occurrence of patterns in **piecewise functions**.

Analogous to Figures 1.1–1.4, you will analyze distinct patterns (sequences) that originate from various arrangements of geometrical figures. Similar to Figures 1.5–1.7, you will decipher and formulate contrasting piecewise functions that are composed of horizontal lines and diagonal lines with a positive and with a negative slope on specific restricted intervals.

Next, let's emerge with three diagrams that examine a variety of geometrical configurations and pose the three related questions that ignite your curiosity on extraction about sequences and their formulations. For instance, in Figure 2.1, what patterns can emanate from the succeeding arrangement of equilateral triangles at the same scale?

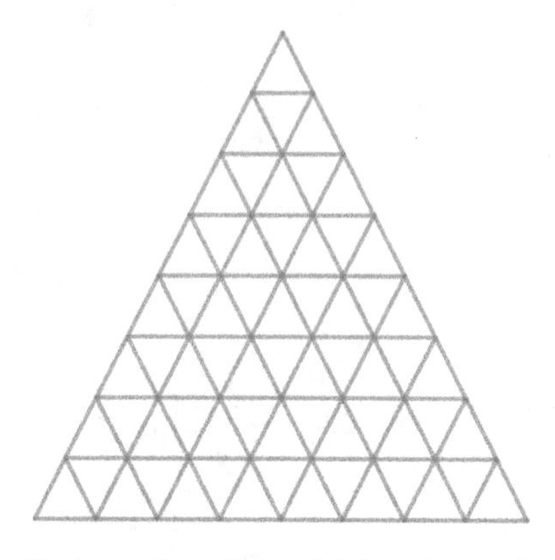

Figure 2.1: System of equilateral triangles at the same scale.

Note that the equilateral triangles in Figure 2.1 mimic the construction of right triangles in Figure 1.1 that add consecutive positive odd integers starting with 1. The upcoming diagram in Figure 2.2, describes the distinct patterns of shrinking squares that arise by inscribing a smaller square (blue square) inside a larger square (green square).

Figure 2.2: System of diminishing squares.

Observe that the shrinking squares in Figure 2.2 mirror the structure of diminishing equilateral triangles in Figure 1.3 and paper folding in Figures 1.8 and 1.9. These will be formulated by a geometrical sequence or pattern. What distinct pattern can you detect in the corresponding geometrical fractal system of diminishing triangles displayed in Figure 2.3?

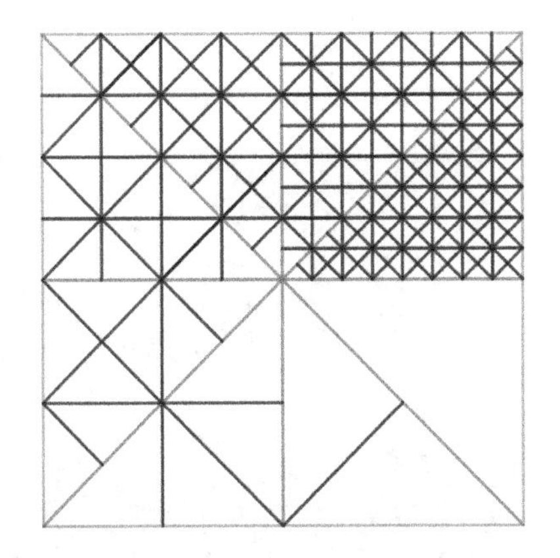

Figure 2.3: System of diminishing 45-45-90 triangles.

In Figure 2.3, the number of triangles in each quadrant increases starting with the fourth quadrant while the size of the right triangles diminishes. Similar occurrence in Figure 1.4 is described by the **geometric summation** that determines the number of triangles.

2.1 Patterns at Same Scale

In this section, you will focus on repetitions of patterns at the same scale and trace and formulate their unique attributes. You can observe repetition of patterns at the same scale in the architectural style of buildings, design of bridges, sidewalks and in the formations of water fountains. The following photograph captures the recurrent aqua patterns of the evening water fountain in Druskininkai, Lithuania:

The upcoming photograph presents a replicated system of arcs of the railroad bridge across the Daugava River in Riga, Latvia:

You can also note repetition of patterns at the same scale while spending time outdoors such as the configurations of waves, mountains, clouds, trees, dunes, waterfalls, etc. The next photograph resembles the repetitive mini alpine formations of the dunes carved by the winds along the Baltic seashores in Liepaja, Latvia:

You can also find the duplicated cascading patterns in Franconia Notch State Park in New Hampshire as shown in the corresponding photograph:

The next Section 2.1.1 will examine repetition of patterns at same scale that emerges in piecewise functions.

2.1.1 *Piecewise Functions*

In Figures 1.5–1.7, you encountered piecewise functions with replicated patterns at the same scale. Analogous to Figure 1.5, the following sketch in Figure 2.4 resembles a sequence of positive even integers on restricted intervals with length 2.

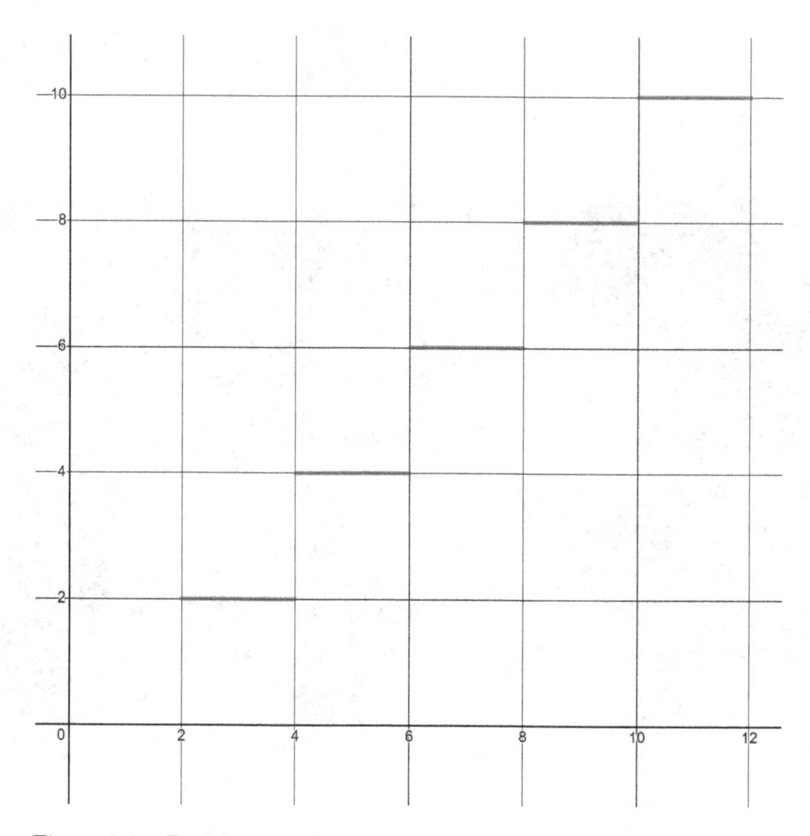

Figure 2.4: Positive even integers on restricted intervals with length 2.

Figure 2.4 commences with $y = 2$ on the interval $(2, 4)$, $y = 4$ on the interval $(4, 6)$, $y = 6$ on the interval $(6, 8)$ and so on. Then for all $n \in \mathbb{N}$, you formulate the corresponding piecewise equation:

$$y = \begin{cases} 2 & \text{if } x \in (2,4) \\ 4 & \text{if } x \in (4,6) \\ 6 & \text{if } x \in (6,8) \\ 8 & \text{if } x \in (8,10) \\ \vdots & \\ 2n & \text{if } x \in (2n, 2n+2) \\ \vdots & \end{cases}$$

The consequent sketch in Figure 2.5 presents an alternating switch of patterns between diagonal lines with a positive slope 1 and horizontal lines on restricted intervals with equal length 1.

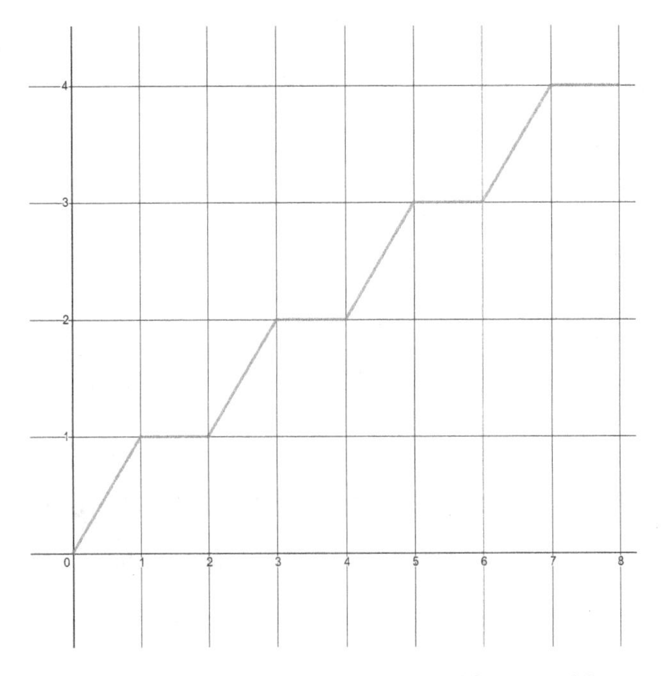

Figure 2.5: System of diagonal lines and horizontal lines.

Figure 2.5 starts off with $y = x$ on the interval $[0, 1]$, $y = 1$ on the interval $[1, 2]$, $y = x - 1$ on the interval $[2, 3]$, $y = 2$ on the interval $[3, 4]$ and so on. Then for all $n \geq 0$, you formulate the corresponding piecewise equation:

$$y = \begin{cases} x & \text{if } x \in [0, 1] \\ 1 & \text{if } x \in [1, 2] \\ x - 1 & \text{if } x \in [2, 3] \\ 2 & \text{if } x \in [3, 4] \\ x - 2 & \text{if } x \in [4, 5] \\ 3 & \text{if } x \in [5, 6] \\ x - 3 & \text{if } x \in [6, 7] \\ 4 & \text{if } x \in [7, 8] \\ \vdots & \\ x - n & \text{if } x \in [2n, 2n + 1] \\ n + 1 & \text{if } x \in [2n + 1, 2n + 2] \\ \vdots & \end{cases} \tag{2.1}$$

Observe that in Figure 2.5 and in (2.1) the slope of all the blue lines is 1. In the meantime, the y-intercepts of the blue lines starting with 0 descend with increments of 1.

In Figure 2.4, the corresponding restricted intervals were of equal length 2 for each piecewise component, while in Figure 2.5 the corresponding restricted intervals were of equal length 1 for each piecewise component.

Next, you will examine a piecewise function where its piecewise components will not have equal length intervals. The succeeding diagram in Figure 2.6 specifies a system of two horizontal lines on restricted intervals with length 1 and with length 3.

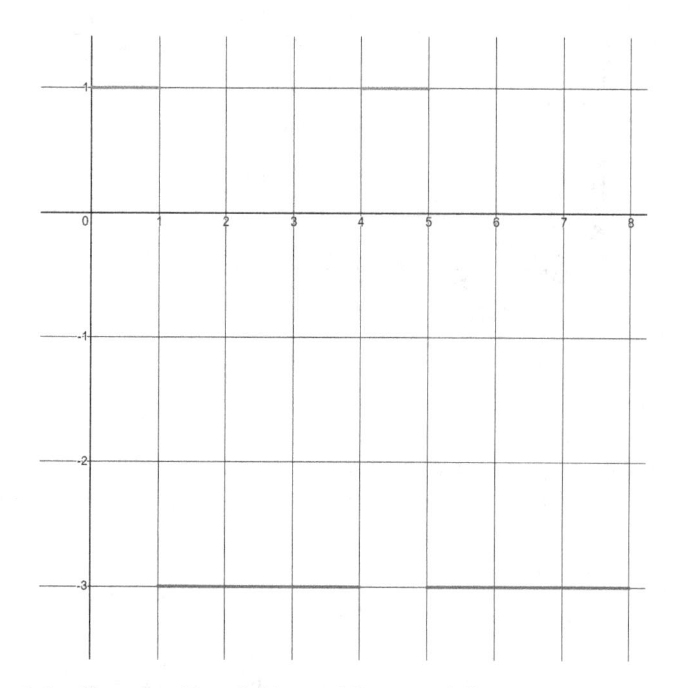

Figure 2.6: System of two horizontal lines on different restricted intervals.

Note that in Figure 2.6, the blue line $y = 1$ is on the restricted interval $(0, 1)$ with length 1, while the green line $y = -3$ is on restricted interval $(1, 4)$ with length 3. Next, you decompose the interval $(0, 4)$ into two disjoint intervals as: $(0, 1) \cup (1, 4)$.

Analogously, the blue line $y = 1$ is on the restricted interval $(4, 5)$ with length 1, while the green line $y = -3$ is on restricted interval $(5, 8)$ with length 3. You decompose the intervals $(4, 8)$ into two disjoint intervals as: $(4, 5) \cup (5, 8)$.

Hence, notice that the intervals of length 4 that are decomposed into intervals with length 1 and length 3 and have $1 : 3$ ratio. Then for all $n \geq 0$, you formulate the corresponding piecewise equation:

$$y = \begin{cases} 1 & \text{if } x \in (0,1) \\ -3 & \text{if } x \in (1,4) \\ 1 & \text{if } x \in (4,5) \\ -3 & \text{if } x \in (5,8) \\ 1 & \text{if } x \in (8,9) \\ -3 & \text{if } x \in (9,12) \\ \vdots & \\ 1 & \text{if } x \in (4n, 4n+1) \\ -3 & \text{if } x \in (4n+1, 4n+4) \\ \vdots & \end{cases}$$

The next section will direct your focus on configurations that portray repetition of patterns at the same scale of various geometrical configurations such as systems of squares and triangles.

2.1.2 *Geometrical Structures*

Figures 1.1 and 1.2 in Chapter 1 render replicated shapes at the same scale. You will proceed with further supplemental examples that resemble reoccurring formations and their specific patterns. You will use horizontal lines, vertical lines and diagonal lines on corresponding restricted intervals to construct and describe particular systems of squares and triangles. Parallel to Figure 1.10, the next two figures analyze a system of squares together with (1.37). The first example formulates a pyramid-shaped system of squares.

The corresponding sketch in Figure 2.7 resembles eight rows of a pyramid-shaped system of squares.

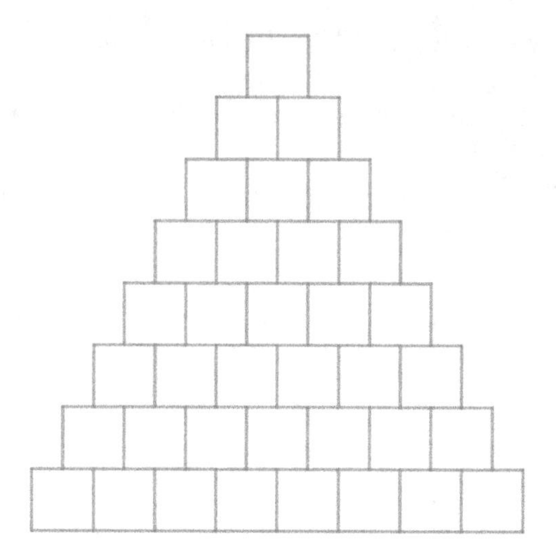

Figure 2.7: A pyramid-shaped system of squares.

Note that Figure 2.7 has eight rows of squares, where the first row has one square, the second row has two squares, the third row has three squares, the fourth row has four squares and so on. You then procure the cognate pattern that cites all the consecutive positive integers between 1 and 8:

$$1,\ 2,\ 3,\ 4,\ 5,\ 6,\ 7,\ 8 = \{i\}_{i=1}^{8}. \tag{2.2}$$

Parallel to Figure 1.10, you obtain the corresponding sum by adding all the squares in each row:

$$1 + 2 + 3 + 4 + 5 + 6 + 7 + 8 = \sum_{i=1}^{8} i. \tag{2.3}$$

For all $n \in \mathbb{N}$, (2.3) extends to the following summation that adds all the consecutive positive integers starting with 1:

$$1 + 2 + 3 + \cdots + (n-1) + n = \sum_{i=1}^{n} i = \frac{n[n+1]}{2}. \tag{2.4}$$

You will prove (2.4) by using the **proof by induction** method in Chapter 4.

Similar to Figures 1.10 and 2.7, Figure 2.8 portrays a square as a system of 36 squares assembled in either six rows, six columns or 11 diagonals.

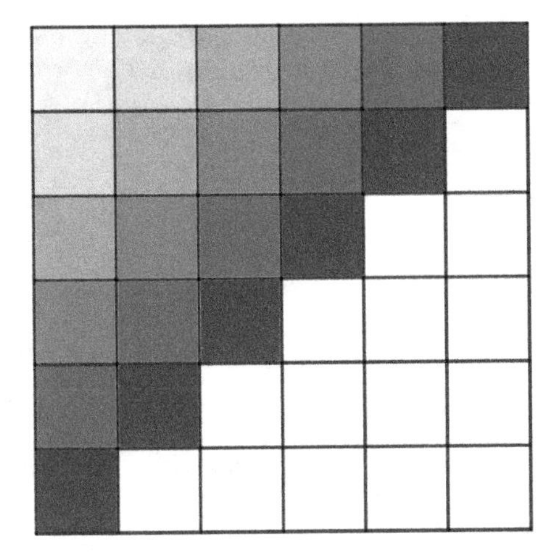

Figure 2.8: A 6×6 square resembled with six blue diagonals.

In Figure 2.8, you commence from the light-blue square in the upper-left corner and include all the blue diagonals toward to the square's principal dark-blue diagonal. You then obtain the corresponding pattern:

$$1,\ 2,\ 3,\ 4,\ 5,\ 6 \ = \ \{i\}_{i=1}^{6}.$$

Analogous to Figures 1.10 and 2.7, you acquire the following sum by adding all the squares in the blue diagonals:

$$1 + 2 + 3 + 4 + 5 + 6 = \sum_{i=1}^{6} i. \tag{2.5}$$

Note that (2.5) also extends to (2.4). The cognate system of horizontal lines and vertical lines on the corresponding restricted intervals sketches the assigned arrangement of squares in Figure 2.8:

$$y = \begin{cases} 3 & \text{for } x \in [-3,3] \\ 2 & \text{for } x \in [-3,3] \\ 1 & \text{for } x \in [-3,3] \\ 0 & \text{for } x \in [-3,3] \\ -1 & \text{for } x \in [-3,3] \\ -2 & \text{for } x \in [-3,3] \\ -3 & \text{for } x \in [-3,3] \end{cases} \qquad x = \begin{cases} 3 & \text{for } y \in [-3,3] \\ 2 & \text{for } y \in [-3,3] \\ 1 & \text{for } y \in [-3,3] \\ 0 & \text{for } y \in [-3,3] \\ -1 & \text{for } y \in [-3,3] \\ -2 & \text{for } y \in [-3,3] \\ -3 & \text{for } y \in [-3,3] \end{cases}$$

Analogous to Figures 1.1 and 2.1, Figure 2.9 describes a system of equilateral triangles assembled with six rows.

Figure 2.9: A system of equilateral triangles at same scale with six rows.

Akin to Figure 1.1, Figure 2.9 has six rows. The first row has one triangle, the second row has three triangles, the third row has five triangles, the fourth row has seven triangles and so on. Hence you see that each row has an odd number of triangles and traces the following pattern that describes the consecutive positive odd integers starting with 1:

$$1,\ 3,\ 5,\ 7,\ 9,\ 11 \ = \ \{(2i-1)\}_{i=1}^{6}.$$

Using (1.16) you will determine the total number of triangles by adding all the triangles in each row and obtain the corresponding sum:

$$1 + 3 + 5 + 7 + 9 + 11 = \sum_{i=1}^{6} (2i-1) = 36 = 6^2. \qquad (2.6)$$

For all $n \in \mathbb{N}$, (2.6) extends to the corresponding summation that adds all the consecutive positive odd integers starting with 1:

$$1 + 3 + 5 + \cdots + (2n-3) + (2n-1) = \sum_{i=1}^{n} (2i-1) = n^2. \ (2.7)$$

You will prove (2.7) using the **proof by induction** technique in Chapter 4. Sketching the decomposed arrangement of right triangles in Figure 2.9 will require a system of horizontal lines, diagonal lines with a positive slope and diagonal lines with a negative slope on various restricted intervals. The diagonal lines will be a family of parallel lines with the same slope but different x-intercepts and y-intercepts. This will be left as an end of chapter exercise.

The succeeding Section 2.2 will formulate **fractals** or replicated geometrical patterns at different scales (at diminishing scales). You will perceive new patterns that govern the specific geometrical arrangements and compare the similarities and differences to what you encountered in this section.

2.2 Patterns at Different Scales

In Chapter 1, you defined a **fractal** as the repetition of patterns at different scales and examined contrasting examples. This section's goals are to widen your horizons on geometrical fractals as you encountered in Figures 1.3, 1.4, 1.8 and 1.9 in Chapter 1 and discover and formulate their unique traits and associated patterns.

For instance, you can observe repetition of patterns at different scales in the architectural styles of buildings. The succeeding photograph captures replicated architectural patterns at different scales of an orthodox monastery, where the largest pattern emerges at the very top:

You can also discover repetitions of patterns at different scales while spending time outdoors such as the coastal formations, the clouds' shapes, the alpine ridges and other fractal-type arrangements. While flying over the Aegean Sea, the corresponding photograph detects the patterns of the coastal formations of the Aegean Islands at diminishing scales starting with the largest island:

The upcoming photograph resembles a system of pyramid-shaped mountains and their alpine reflection ranging in different altitudes and scales in Jasper National Park in the Canadian Rockies:

The consequent photograph traces a system of clouds and their diminishing miens along the Gulf of Riga's shores in Latvia:

It is interesting to note that the clouds' shapes gradually abate as they move further out into the sea. Sometimes the opposite phenomena occurs when the smaller clouds are closer to the shore and the larger clouds are further out in the sea. Next, you will advance with the deeper study of geometrical fractals and their distinct characteristics and formulations.

2.2.1 *Diminishing Geometrical Patterns*

You encountered several examples that render geometrical configurations at diminishing scales such as rectangular and triangular paper folding in Figures 1.8 and 1.9, which are formulated by the cognate **finite geometric sequence**:

$$\left(\frac{1}{2}\right)^0, \left(\frac{1}{2}\right)^1, \left(\frac{1}{2}\right)^2, \left(\frac{1}{2}\right)^3, \left(\frac{1}{2}\right)^4, \left(\frac{1}{2}\right)^5, \left(\frac{1}{2}\right)^6, \left(\frac{1}{2}\right)^7, \left(\frac{1}{2}\right)^8 = \left\{\left(\frac{1}{2}\right)^i\right\}_{i=0}^{8}$$

which then extends to the corresponding **finite geometric sequence:**

$$\{a \cdot r^i\}_{i=0}^{n} = a \cdot r^0 \,, \; a \cdot r^1 \,, \; a \cdot r^2 \,, \; a \cdot r^3 \,, \; \dots \,, \; a \cdot r^n \,. \qquad (2.8)$$

This section's goals are to enhance your knowledge on the construction of geometrical fractals and their cognate patterns that replicate at diminishing scales. You will compare the similarities and differences with the patterns acquired in Chapter 1. You will examine repetitions of patterns at diminishing scales and formulate them by a **geometric sequence.**

Parallel to Figures 1.3, 1.4, and 2.2, Figure 2.10 presents a system of diminishing blue 45-45-90 triangles embedded inside the principle red 45-45-90 triangle.

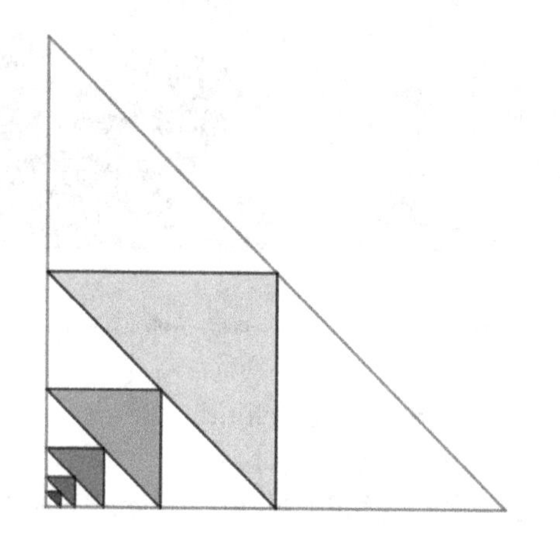

Figure 2.10: System of 45-45-90 triangles.

In Figure 2.10, you construct the sequential blue triangle by extending a horizontal line and a vertical line from the mid-point of the diagonal of the primary red triangle. You continue with the same algorithm by extending a horizontal line and a vertical line from the mid-point of the diagonal of the largest blue triangle and sequentially with respect to all other blue triangle. You then obtain a system of diminishing blue 45-45-90 triangles.

Now suppose that the length of the horizontal and the vertical side of the primary red triangle is 2, then by applying the Pythagorean Theorem and the properties of 45-45-90 triangles, the length of the horizontal and the vertical side of the sequential blue triangle becomes 1. Then for all $n \geq 0$, you procure the corresponding **geometric sequence** that formulates the sequence of diminishing 45-45-90 triangles depicted in Figure 2.10 as:

$$2,\ 1,\ \frac{1}{2},\ \frac{1}{4},\ \frac{1}{8},\ \frac{1}{16},\ \ldots,\ 2^{1-n},\ \ldots. \tag{2.9}$$

Next observe that the area of the primary red triangle is 2 and the area of the sequential blue triangle is $\frac{1}{2}$. Hence for all $n \geq 0$ you procure the following **geometric sequence** that formulates the sequence of areas in Figure 2.10:

$$2,\ \frac{1}{2},\ \frac{1}{8},\ \frac{1}{32},\ \frac{1}{128},\ \ldots,\ 2^{1-2n},\ \ldots.$$

The corresponding version of Figure 2.11 addresses the question regarding the total number of blue 45-45-90 triangles inside the primary 45-45-90 triangle:

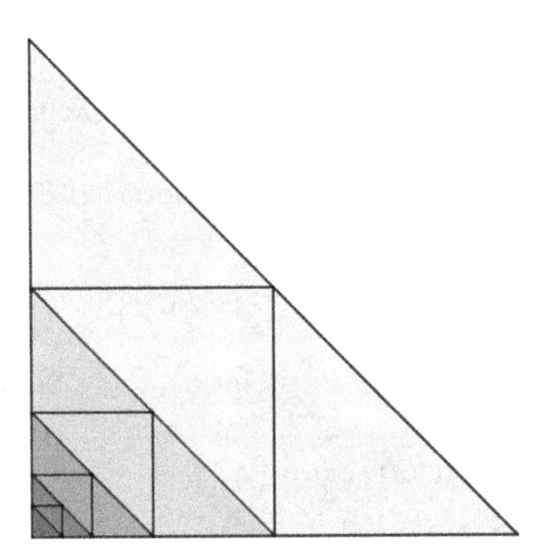

This will be formulated by a linear sequence and will be left as an end of chapter exercise. Similar examples will be done in the latter part of this section.

Akin to Figures 1.3, 1.4, and 2.10, Figure 2.11 examines a system of diminishing squares inscribed inside the principle green square.

Figure 2.11: System of diminishing squares.

In Figure 2.11, you commence with the primary green square and construct the sequential blue square by connecting the diagonal blue lines between the mid-points of the primary green square's sides. The consequent green square is designed by connecting the horizontal and vertical green lines between the mid-points of the blue square's sides and so on. Now suppose that the length of each side of the primary green square is 2. By applying the Pythagorean Theorem and the properties of 45-45-90 triangles, the length of each side of the neighboring blue square becomes $\sqrt{2}$. Hence for all $n \geq 0$, you acquire the following **geometric sequence** that describes the length of sides of the squares:

$$2,\ \sqrt{2},\ 1,\ \frac{1}{\sqrt{2}},\ \frac{1}{2},\ \frac{1}{2\sqrt{2}},\ \ldots,\ \left(\sqrt{2}\right)^{2-n},\ \ldots. \tag{2.10}$$

Note that the green terms of (2.10) present the length of each green square, while the blue terms of (2.10) trace the length of each blue square. You can also reformulate (2.10) as a **piecewise sequence** that consists of two geometric sequences.

Next, note that the area of the primary green square is 4 and the area of the sequential blue square is 2. Hence, for all $n \geq 0$ you procure the following **geometric sequence** that describes the sequence of areas:

$$4,\ 2,\ 1,\ \frac{1}{2},\ \frac{1}{4},\ \frac{1}{8},\ldots,\ 2^{2-n}.$$

Sketching the assigned system of squares in Figure 2.11 will require the use of parallel horizontal, vertical and diagonal lines on the corresponding restricted intervals. In fact, the green squares are depicted by a system of parallel horizontal and vertical lines on the corresponding restricted intervals, while the blue squares are described by a system of parallel diagonal lines with a positive slope 1 and negative slope -1 on the corresponding restricted intervals. This will be left as end of chapter exercises.

Furthermore, by connecting the mid-points of the primary green square in Figure 2.11 you obtain four triangles in between the green and the blue triangles. We then obtain four triangles every time you embed a smaller square inside a bigger square.

Analogous to Figures 2.10 and 2.11, Figure 2.12 mimics a system of diminishing blue and green equilateral triangles.

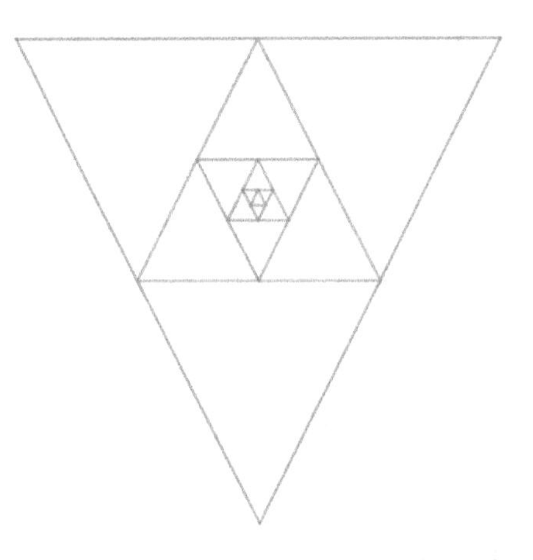

Figure 2.12: System of diminishing equilateral triangles.

In Figure 2.12, you assemble the first green triangle by connecting the green lines between the mid-points of the primary blue triangle and so on. Analogous to Figure 2.11, the sides and the areas of the triangles will be characterized by a geometric sequence.

Suppose that the length of each side the largest blue triangle is 2. Then the length of each side of the consequent green triangle is 1 as it is half the length of the blue triangle and so on. Therefore, for all $n \geq 0$ you acquire the following **geometric sequence** that depicts the length of the sides of all the triangles starting with the primary blue triangle:

$$2,\ 1,\ \frac{1}{2},\ \frac{1}{2^2},\ \frac{1}{2^3},\ \frac{1}{2^4},\ \ldots,\ \left(\frac{1}{2}\right)^{n-1}. \tag{2.11}$$

Now, note that using the properties of 30-60-90 triangles, the area of the primary blue triangle is $\sqrt{3}$ and the area of the adjacent green triangle is $\frac{\sqrt{3}}{4}$. Hence, for all $n \geq 0$, you procure the following **geometric sequence** that describes the sequence of areas:

$$\sqrt{3}, \frac{\sqrt{3}}{4}, \frac{\sqrt{3}}{4^2}, \frac{\sqrt{3}}{4^3}, \frac{\sqrt{3}}{4^4}, \frac{\sqrt{3}}{4^5}, \ldots, \frac{\sqrt{3}}{4^n}.$$

Next, you will determine the number of 45-45-90 triangles at different scales. Analogous to Figure 1.1, Figure 2.13 decomposes the primary red 45-45-90 triangle into six rows of 45-45-90 triangles.

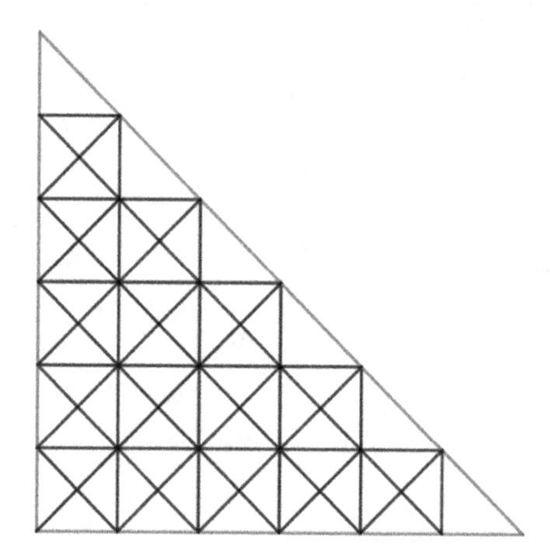

Figure 2.13: An arrangement of 45-45-90 triangles.

Note that Figure 2.13 has six rows where the first row has one triangle, the second row has five triangles, the third row has nine triangles and so on. You transition from row to row by adding four more triangles starting with one triangle in the first row. Therefore, each row has an odd number of triangles described by the following **linear sequence**:

$$1,\ 5,\ 9,\ 13,\ 17,\ 21 = \{4i - 3\}_{i=1}^{6}.$$

The total number of triangles inscribed inside the principal red triangle in Figure 2.13 is specified by the following sum:

$$1 + 5 + 9 + 13 + 17 + 21 = \sum_{i=1}^{6} (4i - 3). \qquad (2.12)$$

Observe that for all $n \in \mathbb{N}$, (2.12) extends to the following sum that adds the corresponding odd integers starting with 1 which differ by 4:

$$1 + 5 + 9 + 13 + \cdots + (4n-3) \;=\; \sum_{i=1}^{n} (4i-3) \;=\; n\cdot[2n-1]. \quad (2.13)$$

You will derive (2.13) and analogous formulas and prove them by applying the **proof by induction** method in Chapter 4.

Your next aim is to determine the total number of red triangles presented in Figure 2.14.

Figure 2.14: System of symmetrical 45-45-90 triangles.

Note that each time you incise a blue square inside a green square and *vice versa* you produce four symmetrical red 45-45-90 triangles. Therefore, for $n \in \mathbb{N}$, the total number of generated triangles mimics the corresponding **linear pattern**:

$$4,\ 8,\ 12,\ 16,\ \ldots,\ 4n \;=\; \{4 \cdot i\}_{i=1}^{n}.$$

You will discover additional linear patterns that arise and resemble similar geometrical configurations.

Next, you will determine the number of red 30-60-90 triangles illustrated in Figure 2.15.

Figure 2.15: System of red symmetrical equilateral triangles.

In Figure 2.15, by linking the lines between the three mid-points of the principal blue triangle you construct three red equilateral sub-triangles. In this case, three equilateral triangles arise when you inscribe a smaller triangle insider a larger triangle. This is described by a specific linear sequence that cites 1 more than a multiple of 3. Thus, for all $n \in \mathbb{N}$, the following **linear sequence** determines the total number of red triangles including the smallest dark red triangle in the very center:

$$1,\ 4,\ 7,\ 10,\ \ldots,\ 3n+1 \ = \ \{3i+1\}_{i=0}^{n}.$$

Similar to Figures 1.4, 1.8 and 1.9, Figure 2.16 paints a fractal system of a square with 45-45-90 triangles at diminishing scales.

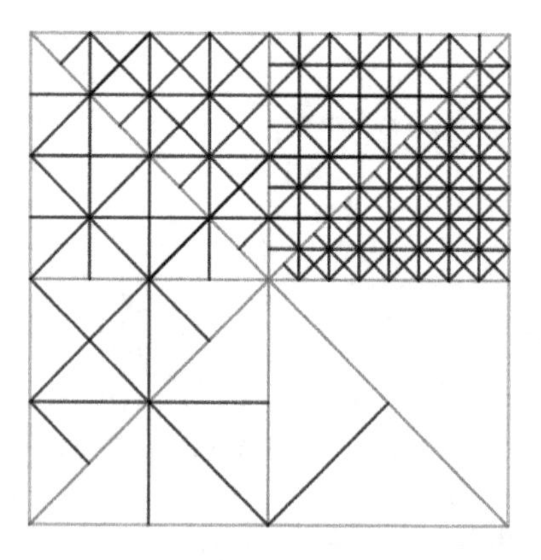

Figure 2.16: System of diminishing 45-45-90 triangles.

The objective is to determine the total number of diminishing 45-45-90 triangles inscribed inside the square. This will guide you to a geometric summation. In Figure 2.16, starting from the largest 45-45-90 triangle, the exact number of 45-45-90 triangles that are inserted inside the blue square is formulated by the following **geometric sum**:

$$\begin{aligned} & 1 + 2 + 4 + 8 + 16 + 32 + 64 + 128 \\ = & \ 2^0 + 2^1 + 2^2 + 2^3 + 2^4 + 2^5 + 2^6 + 2^7 \\ = & \sum_{i=0}^{7} 2^i = 2^8 - 1. \end{aligned} \tag{2.14}$$

Observe (2.14) adds 8 terms as the blue square is decomposed into 8 primary triangular regions (emphasized by the red diagonals, horizontal and vertical lines) and into 8 categories of 45-45-90 triangles. Hence, for all $n \in \mathbb{N}$, (2.14) extends to the corresponding **geometric summation**:

$$\begin{aligned} & a \cdot r^0 + a \cdot r^1 + a \cdot r^2 + a \cdot r^3 + \cdots + a \cdot r^n \\ & = \sum_{i=0}^{n} a \cdot r^i = \frac{a[1 - r^{n+1}]}{1 - r}. \end{aligned} \tag{2.15}$$

Note that (2.15) has $n + 1$ terms added and $r \neq 1$. You will verify (2.15) by using the **proof by induction** technique in Chapter 4.

Figure 2.17 resembles a system of diminishing blue squares and green circles.

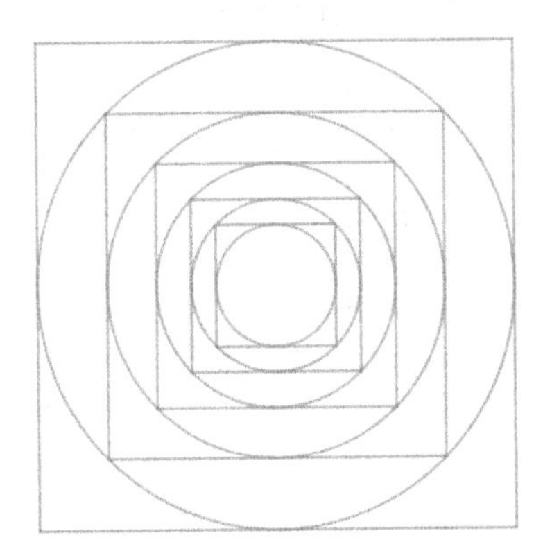

Figure 2.17: System of diminishing squares and circles.

You will describe the dynamics as a piecewise geometric sequence as we see alterations between squares and circles. Figure 2.17 commences by inscribing a green circle inside the main blue square with side 2, whose corresponding area is 4. Note that the circle's diameter is also 2 (radius 1). The diagonal of the next blue square inscribed the green circle is 2 and hence the length of the blue square is $\sqrt{2}$, whose corresponding area is 2. Analogous to Figure 2.11, for all $n \geq 0$ the following **geometric sequence** describes the length of the blue squares starting with the primary square:

$$2,\ \sqrt{2},\ 1,\ \frac{1}{\sqrt{2}},\ \frac{1}{2},\ \frac{1}{2\sqrt{2}},\ \ldots,\ \left(\frac{1}{\sqrt{2}}\right)^{n-2}. \tag{2.16}$$

In addition, for all $n \geq 0$ you obtain the corresponding **geometric sequence** that presents the areas of the blue squares:

$$4,\ 2,\ 1,\ \frac{1}{2},\ \ldots,\ \left(\frac{1}{2}\right)^{n-2}.$$

Then for all $n \geq 0$, you produce the corresponding **geometric sequence** that depicts the radii of the green circles commencing with the largest green circle:

$$1,\ \frac{1}{\sqrt{2}},\ \frac{1}{2},\ \frac{1}{2\sqrt{2}},\ \ldots,\ \left(\frac{1}{\sqrt{2}}\right)^{n}. \tag{2.17}$$

Finally for all for $n \geq 0$, you procure the **geometric sequence** that describes the areas of the green circles:

$$\pi,\ \frac{\pi}{2},\ \frac{\pi}{4},\ \frac{\pi}{8},\ \ldots,\ \frac{\pi}{2^n}.$$

To formulate a sequence describing the alternating pattern of the blue squares and the green circles at the same time will require a **piecewise geometric sequence** that will consists of two distinct geometric sequences of sides of blue squares (2.16) and radii of green circles (2.17). This concept will be addressed in thorough details in the latter part of the next section and in Chapter 3 as well.

Figure 2.17 is a special case of an alternating pattern and will direct you to supplemental studies of alternating patterns.

2.3 Alternating and Piecewise Patterns

Figures 1.2, 1.3, 1.6, 1.7, 2.5, 2.6, 2.11, and 2.17 portrayed alternating and piecewise patterns. This section's aims are to study deeper traits of alternating patterns and their distinct attributes. You will analyze meticulous details with formulating **piecewise sequences** that consist of two or more sequences to describe the specific phenomena analogous to Figure 2.17.

The photograph below portrays an example of alternating architectural patterns of Liepaja University's main building depicting the alternating red and white colors and architectural styles and decorations:

You can also detect nature's alternating schemes while traveling and spending time outdoors which could emerge as alternating colors, alternating themes, etc. The next aerial photograph displays the alternating green and white colors of the French Alps:

The succeeding photograph exhibits the alternating sand patterns formed by the waves and the winds:

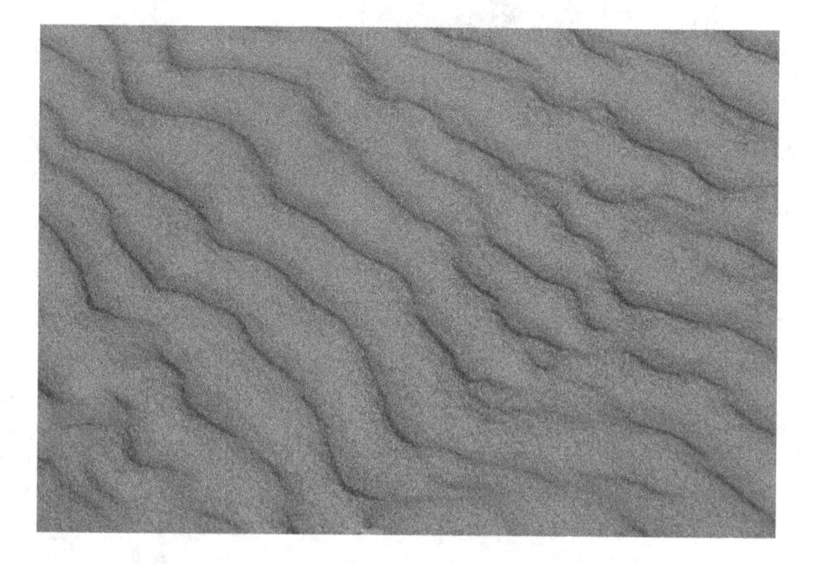

2.3.1 *Alternating Geometrical Patterns*

You encountered several specimens of alternating geometric patterns such as Figure 1.2 portraying alternating right triangles at the same scale. On the other hand, Figure 2.11 presents an alternating formation of diminishing squares. Figure 1.3 describes a system of alternating diminishing equilateral triangles and Figure 2.17 depicts a shift between a square and a circle.

You came across alternating patterns while examining piecewise functions in Figures 1.7, 2.5, and 2.6. Analogous to Figure 2.6, the sketch in Figure 2.18 identifies an alternating system of two horizontal lines on restricted intervals with length 2 and length 1.

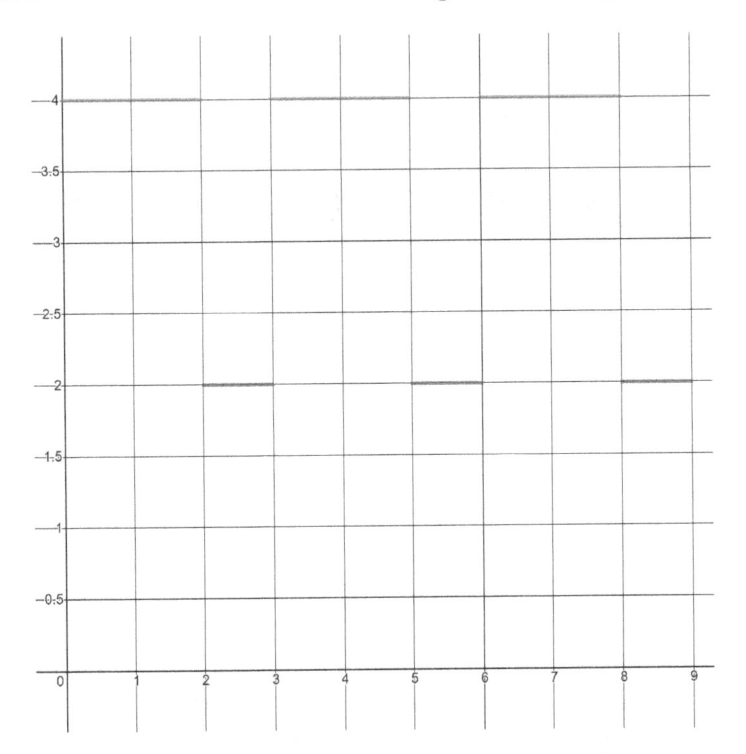

Figure 2.18: Alternating system of two horizontal lines.

In Figure 2.18, the blue line $y = 4$ is on the restricted interval $(0, 2)$ with length 2, while the red line $y = 2$ is on restricted interval $(2, 3)$ with length 1. Next you decompose the interval $(0, 3)$ as: $(0, 2) \cup (2, 3)$.

Analogously, the blue line $y = 4$ is on the restricted interval $(3, 5)$ with length 2, while the red line $y = 2$ is on restricted interval $(5, 6)$ with length 1 and we decompose the intervals $(3, 6)$ into two disjoint intervals as: $(3, 5) \cup (5, 6)$.

Hence the intervals of length 3 that are decomposed into intervals with length 2 and length 1 and have $2 : 1$ ratio. Then for all $n \geq 0$, you formulate the corresponding piecewise function describing the horizontal lines in Figure 2.18:

$$y = \begin{cases} 4 & \text{if } x \in (0, 2) \\ 2 & \text{if } x \in (2, 3) \\ 4 & \text{if } x \in (3, 5) \\ 2 & \text{if } x \in (5, 6) \\ 4 & \text{if } x \in (6, 8) \\ 2 & \text{if } x \in (8, 9) \\ \vdots & \\ 4 & \text{if } x \in (3n, 3n + 2) \\ 2 & \text{if } x \in (3n + 2, 3n + 3) \\ \vdots & \end{cases}$$

Figure 1.2 characterizes the alternating formation of green and blue 45-45-90 triangles. You will revisit some of examples from the previous sections of this Chapter and examine the alternating patterns in further details and express them as **piecewise sequences** that consist of two sequences. The succeeding example explores an alternating system of diminishing squares that you analyzed in Figure 2.11 and resembles the sides and the areas of the green and blue squares as a **piecewise sequence**.

Example 2.1. Recall that Figure 2.11 outlines the configuration of green and blue squares starting with the principal green square with length 2:

By applying the Pythagorean Theorem together and the identities of 45-45-90 triangles, the length of the neighboring blue square becomes $\sqrt{2}$. Hence, for all $n \geq 0$, you procure the following **geometric sequence** that traces the lengths of the squares:

$$2,\ \sqrt{2},\ 1,\ \frac{1}{\sqrt{2}},\ \frac{1}{2},\ \frac{1}{2\sqrt{2}},\ \ldots,\ \left(\frac{1}{\sqrt{2}}\right)^{n-2},\ \ldots. \tag{2.18}$$

Via (2.18), for all $n \geq 0$, you obtain the following **piecewise sequence** that describes the lengths of the green and blue squares:

$$\{l_n\}_{n=0}^{\infty} = \begin{cases} \left(\frac{1}{2}\right)^{\frac{n-2}{2}} & \text{if } n = 0, 2, 4, 6, \ldots, \\ \left(\frac{1}{2}\right)^{\frac{n-2}{2}} & \text{if } n = 1, 3, 5, 7, \ldots. \end{cases} \tag{2.19}$$

The green terms of (2.19) render the length of each green square, while the blue terms of (2.19) characterize the length of each blue square. Recall that the area of the principal green square is 4 and the area of the adjacent blue square is 2. Thus, for all $n \geq 0$, you procure the following **geometric sequence** that describes the sequence of the related areas:

$$4,\ 2,\ 1,\ \frac{1}{2},\ \frac{1}{4},\ \frac{1}{8},\ \ldots,\ \left(\frac{1}{2}\right)^{n-2},\ \ldots. \tag{2.20}$$

Via (2.20), for all $n \geq 0$ you procure the following **piecewise sequence** that describes the areas of the green and blue squares:

$$\{A_n\}_{n=0}^{\infty} = \begin{cases} \left(\frac{1}{2}\right)^{n-2} & \text{if } n = 0, 2, 4, 6, \ldots, \\ \left(\frac{1}{2}\right)^{n-2} & \text{if } n = 1, 3, 5, 7, \ldots. \end{cases}$$

The upcoming example investigates an alternating system of diminishing squares and circles that you examined in Figure 2.17.

Example 2.2. Figure 2.17 commences with inscribing a green circle inside a blue square with length 2 as illustrated below:

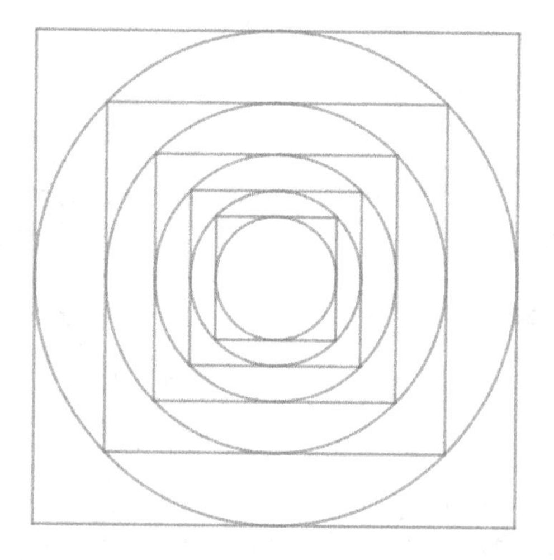

Your aim is formulate the sides of the blue squares, the radii of the green circles as well as their corresponding areas as **piecewise sequences**. By applying the Pythagorean Theorem along with the properties of 45-45-90 triangles, for all $n \geq 0$, you formulate the following **geometric sequence** depicting the lengths of the blue squares:

$$2,\ \sqrt{2},\ 1,\ \frac{1}{\sqrt{2}},\ \frac{1}{2},\ \frac{1}{2\sqrt{2}},\ \ldots,\ \left(\frac{1}{\sqrt{2}}\right)^{n-2},\ \ldots, \tag{2.21}$$

and the corresponding **geometric sequence** rendering the radii of the green circles:

$$1,\ \frac{1}{\sqrt{2}},\ \frac{1}{2},\ \frac{1}{2\sqrt{2}},\ \ldots,\ \left(\frac{1}{\sqrt{2}}\right)^{n},\ \ldots. \tag{2.22}$$

Via (2.21) and (2.22), you formulate the relevant **piecewise sequence**:

$$\{a_n\}_{n=0}^{\infty} = \begin{cases} \left(\frac{1}{\sqrt{2}}\right)^{\frac{n-4}{2}} & \text{if } n = 0, 2, 4, 6, \ldots, \\ \left(\frac{1}{\sqrt{2}}\right)^{\frac{n-1}{2}} & \text{if } n = 1, 3, 5, 7, \ldots. \end{cases}$$

In addition, for all $n \geq 0$, you formulate the following **geometric sequence** describing the areas of the blue squares:

$$4,\ 2,\ 1,\ \frac{1}{2},\ \ldots,\ \left(\frac{1}{2}\right)^{n-2},\ \ldots, \tag{2.23}$$

and the following **geometric sequence** tracing the areas of the green circles:

$$\pi,\ \frac{\pi}{2},\ \frac{\pi}{4},\ \frac{\pi}{8},\ \ldots,\ \frac{\pi}{2^n},\ \ldots. \tag{2.24}$$

Via (2.23) and (2.24), you obtain the corresponding **piecewise sequence**:

$$\{A_n\}_{n=0}^{\infty} = \begin{cases} \left(\frac{1}{2}\right)^{\frac{n-4}{2}} & \text{if } n = 0, 2, 4, 6, \ldots, \\ \pi\left(\frac{1}{2}\right)^{\frac{n-1}{2}} & \text{if } n = 1, 3, 5, 7, \ldots. \end{cases}$$

The upcoming section will conclude the chapter by applying the summation of consecutive integers (3.3) and the summation of consecutive odd integers (3.4) to determine specific areas.

2.4 Summation of Areas

This section's aims are to apply (3.3) and (3.4) to either determine the total area of all the squares and triangles or to determine the area of an individual square and triangle.

Figure 2.19 is a system of pyramid-shaped squares with eight rows. Determine the dimensions of each individual square given that the area of all the squares is 900.

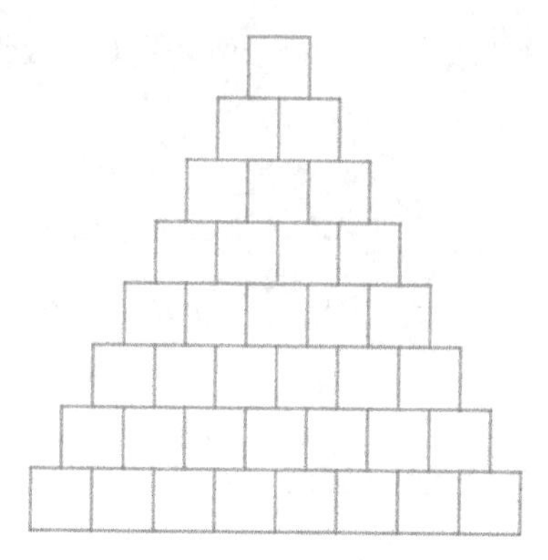

Figure 2.19: System of pyramid-shaped squares at the same scale.

Solution: Using Eq. (3.3), the total number of squares in Figure 2.19 is:

$$\sum_{i=1}^{8} i = \frac{8 \cdot 9}{2} = 36.$$

Hence, the area of each square is the total area divided by the total number of squares:

$$\frac{900}{36} = 25.$$

Thus the dimensions of each square are 5×5.

Figure 2.20 is a system of triangles with eight rows. Suppose that the area of $\triangle ABC$ is 15. Determine the area of $\triangle ADE$.

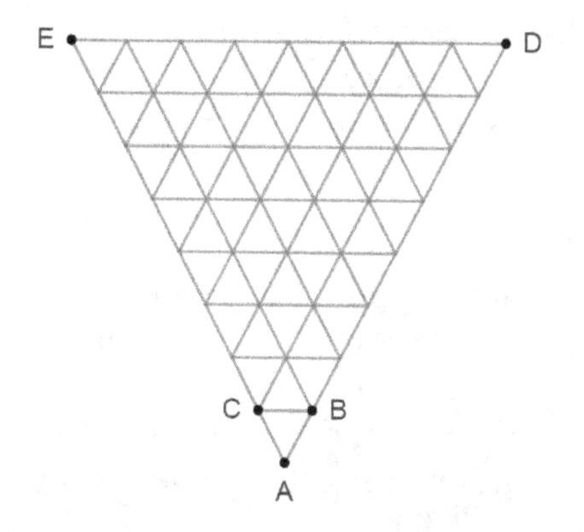

Figure 2.20: System of triangles at the same scale.

Solution: By applying Eq. (3.4), the total number of triangles in Figure 2.20 is:

$$\sum_{i=1}^{8} (2i - 1) = 8^2 = 64.$$

Hence, the area of $\triangle ADE$ is:

$$15 \cdot 64 = 960.$$

Figure 2.21 presents a system of step-shaped squares with five rows. Solve for k when the dimension of each small square is k and the perimeter P (the sum of all the red sides) is equal to the area A of all the squares.

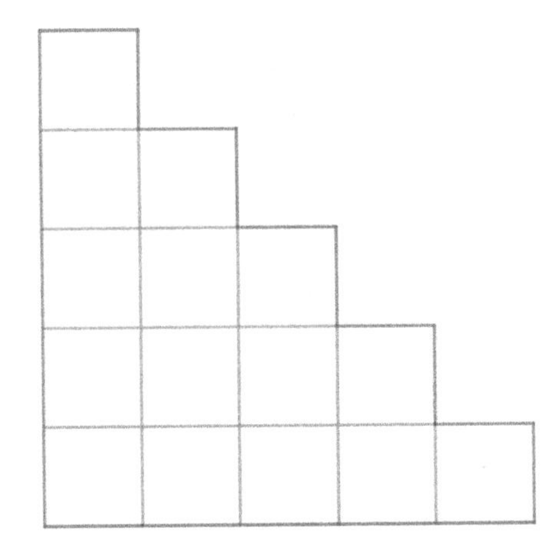

Figure 2.21: System of step-shaped squares at the same scale.

Solution: Via Eq. (2.4), Figure 2.21 has 15 squares and you acquire:

$$A = 15k^2 \quad \text{and} \quad P = 20k.$$

Now set $A = P$ and obtain:

$$15k^2 = 20k,$$

and

$$k = \frac{20}{15} = \frac{4}{3}.$$

Figure 2.21 can be extended to a system of step-shaped squares at the same scale with n rows. This will be left as an end of chapter exercise.

2.5 Exercises

1. Write a **piecewise formula** of the following function:

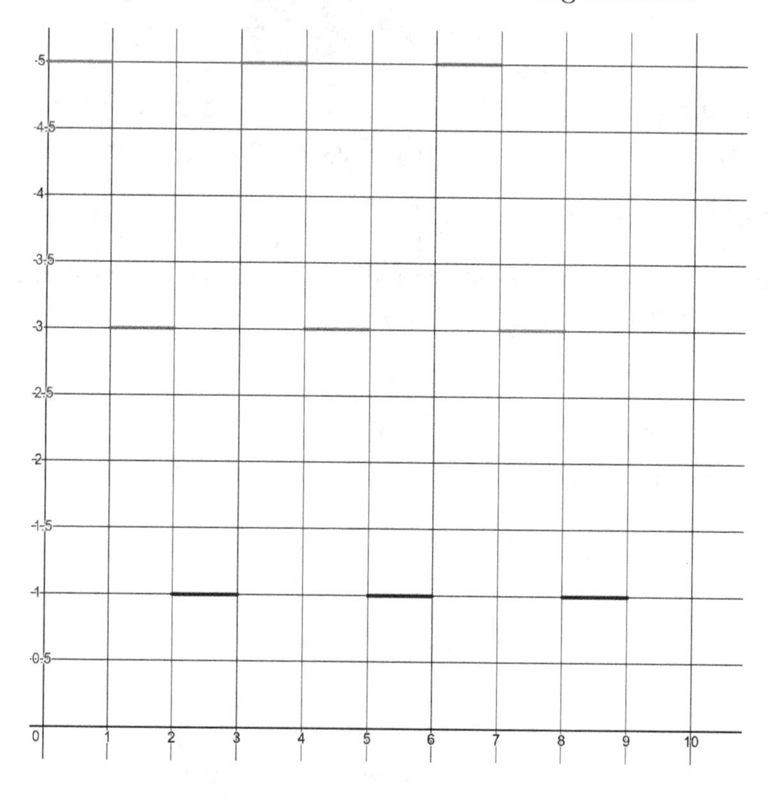

2. Write a **piecewise formula** of the following function:

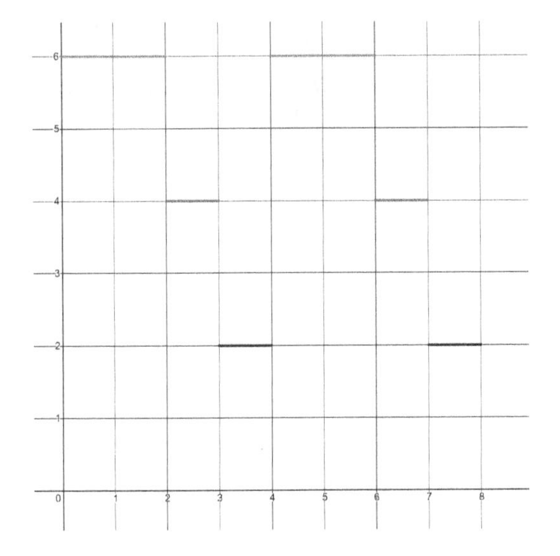

3. Write a **piecewise formula** of the following function:

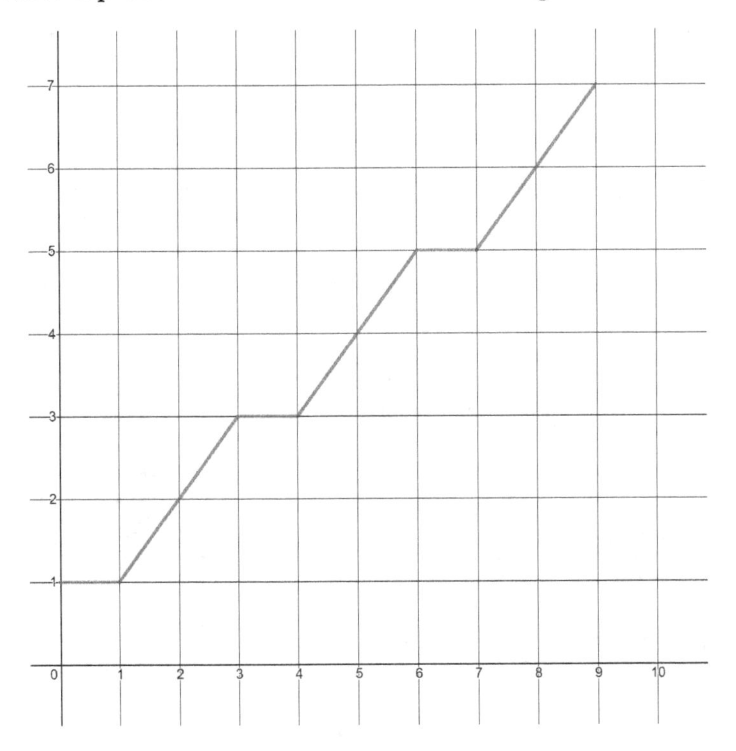

4. Write a **piecewise formula** of the following function:

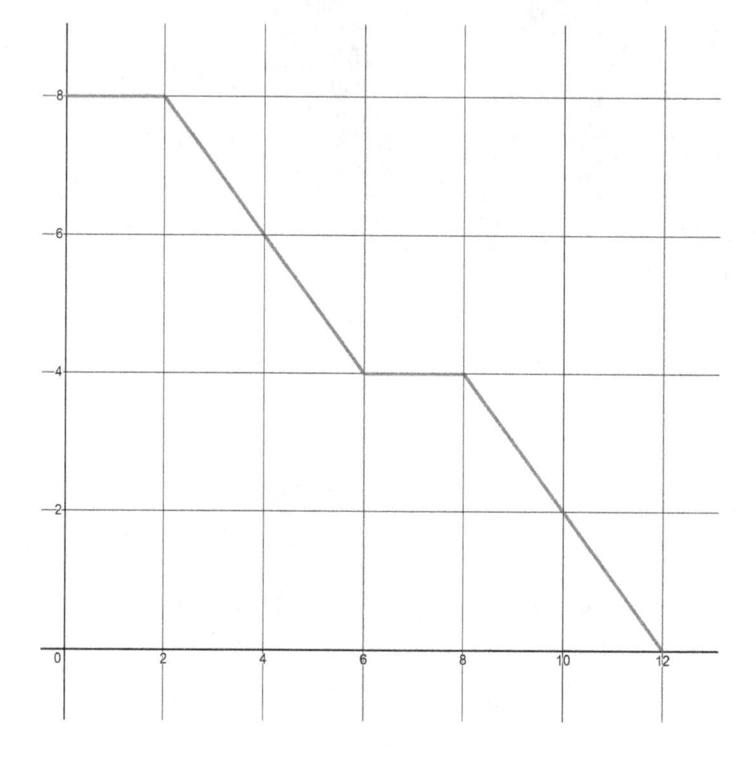

5. Write a **piecewise formula** of the following function:

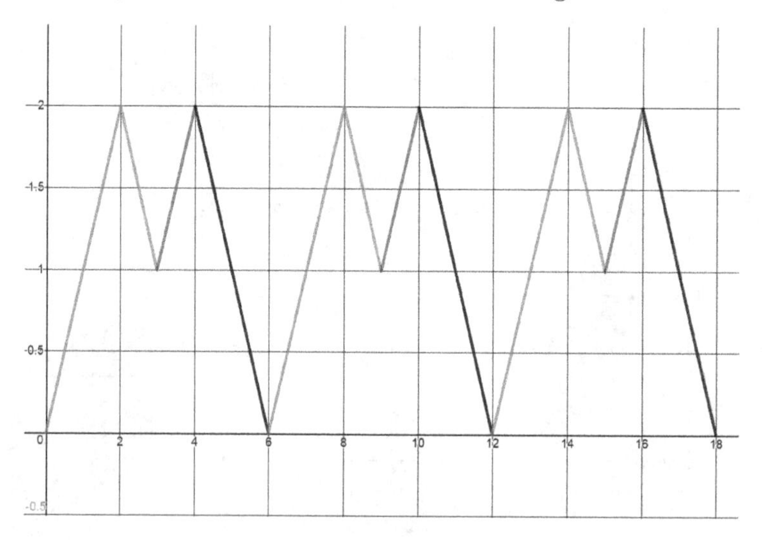

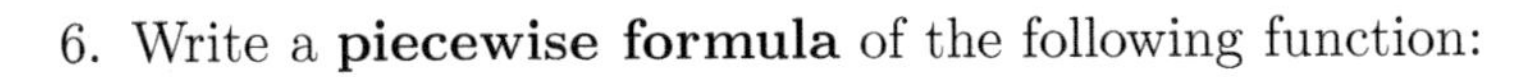

6. Write a **piecewise formula** of the following function:

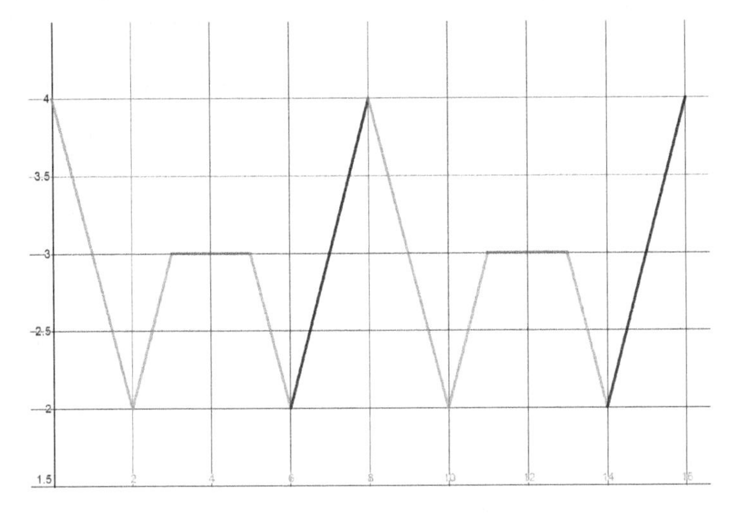

7. Write **piecewise functions** on the cognate restricted intervals that render the corresponding system of shrinking rectangles:

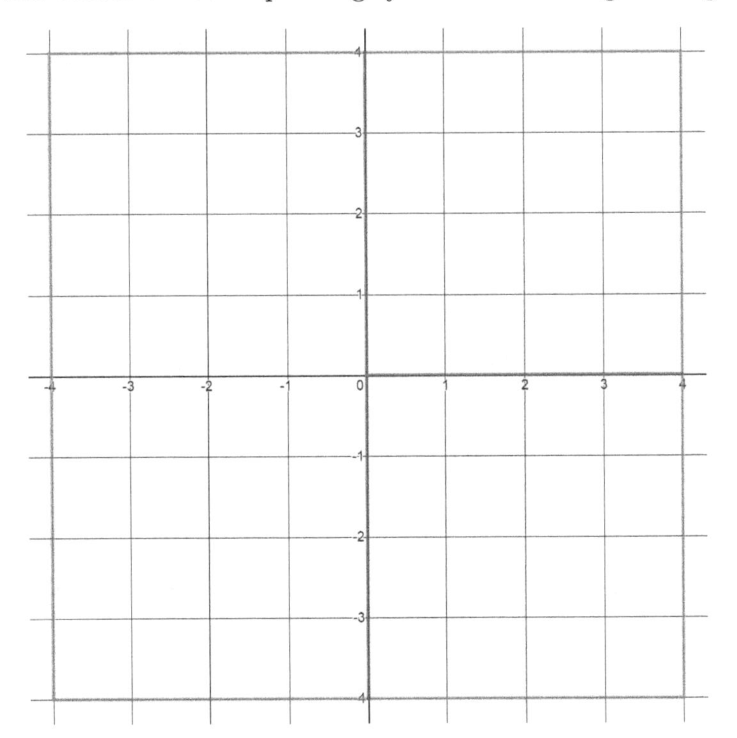

8. Write **piecewise functions** on the cognate restricted intervals that render the corresponding system of shrinking rectangles:

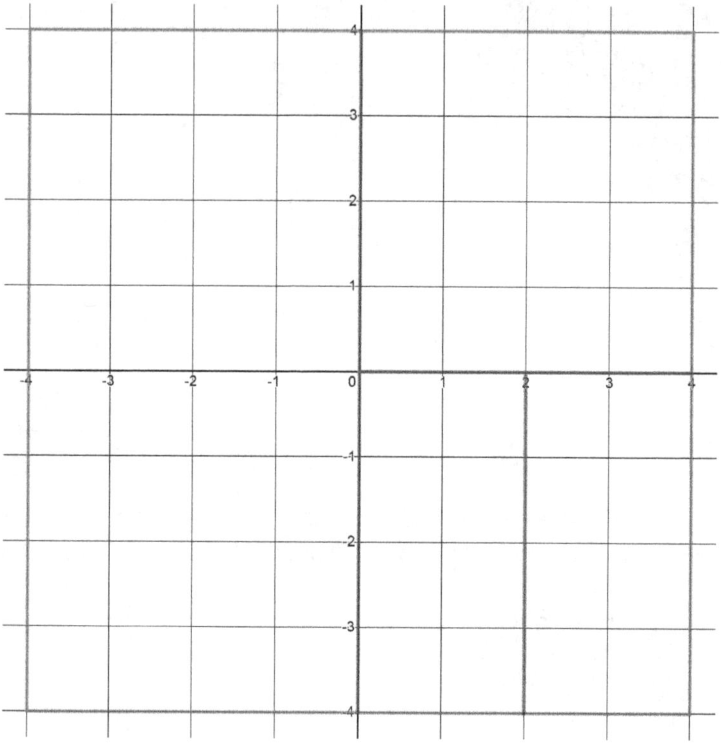

9. Write **piecewise functions** on the cognate restricted intervals that render the corresponding system of shrinking rectangles:

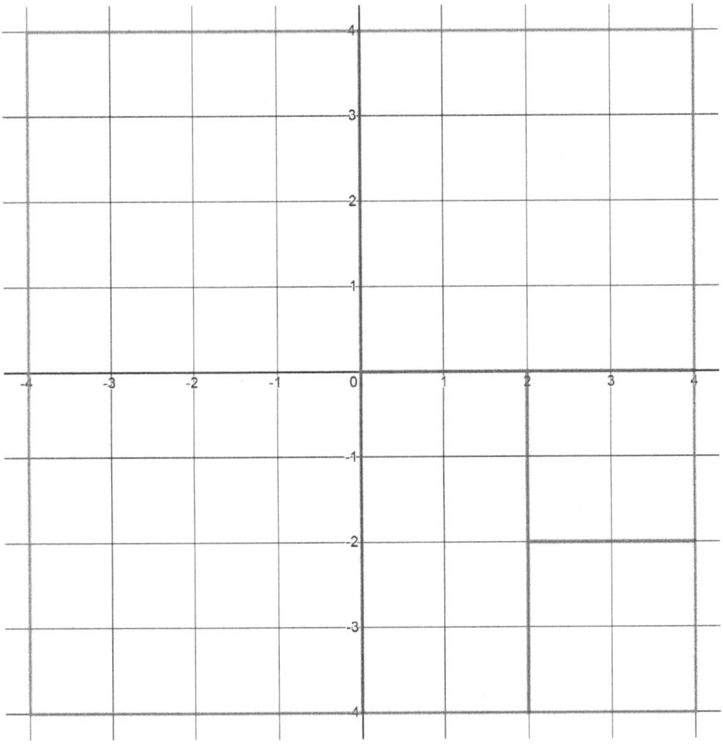

10. Write **piecewise functions** on the cognate restricted intervals that render the corresponding system of shrinking rectangles:

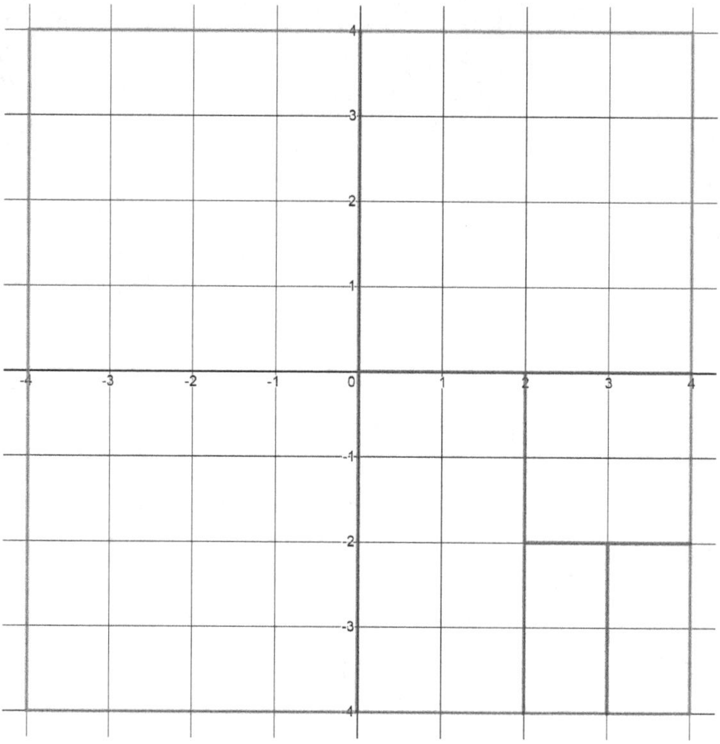

11. Write **piecewise functions** on the cognate restricted intervals that render the corresponding system of shrinking rectangles:

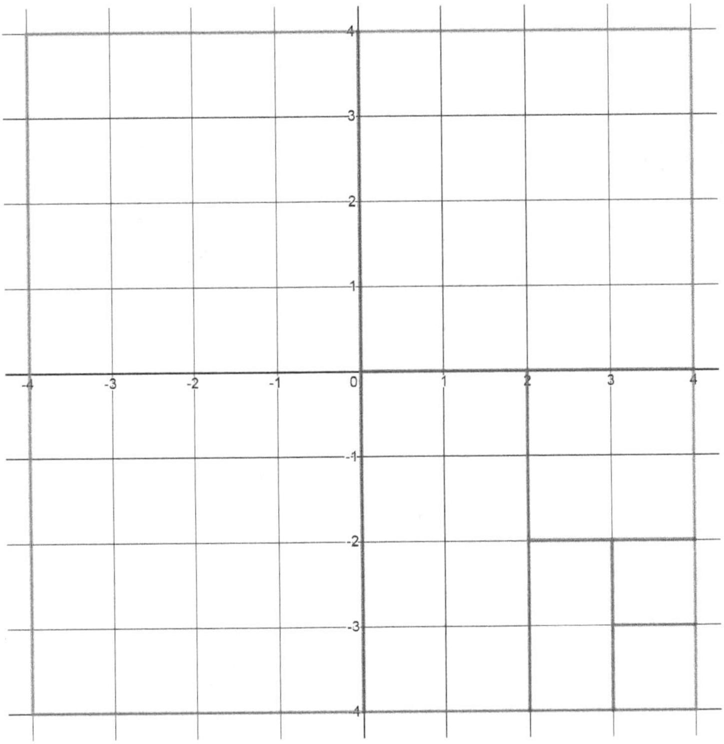

12. Write **piecewise functions** on the cognate restricted intervals that render the corresponding system of shrinking rectangles:

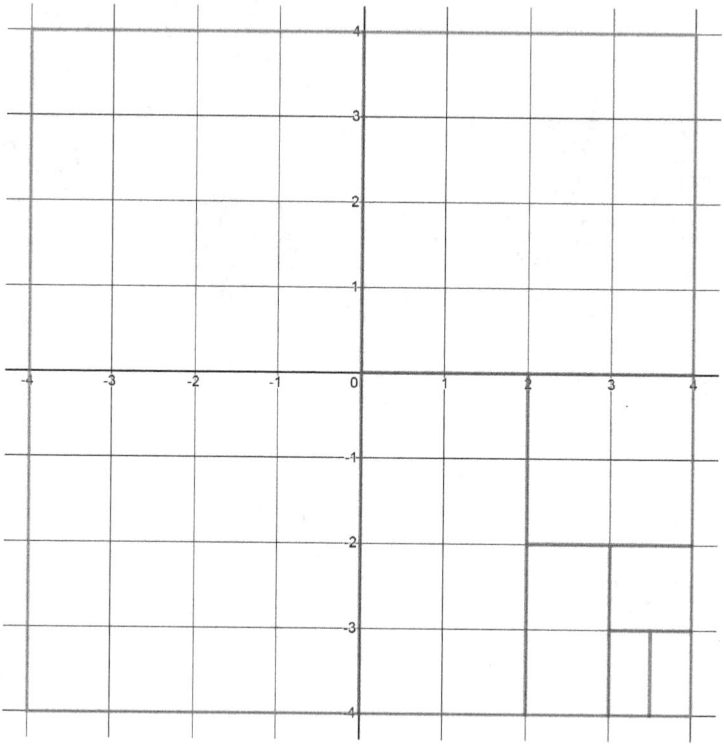

13. Write **piecewise functions** on the cognate restricted intervals that render the corresponding system of right triangles:

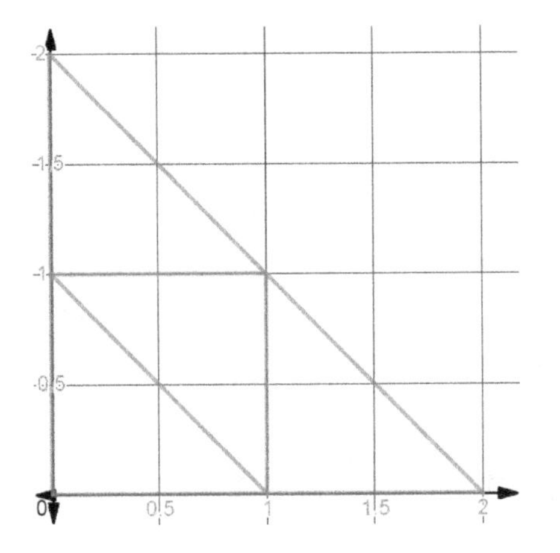

14. Write **piecewise functions** on the cognate restricted intervals that render the corresponding system of right triangles:

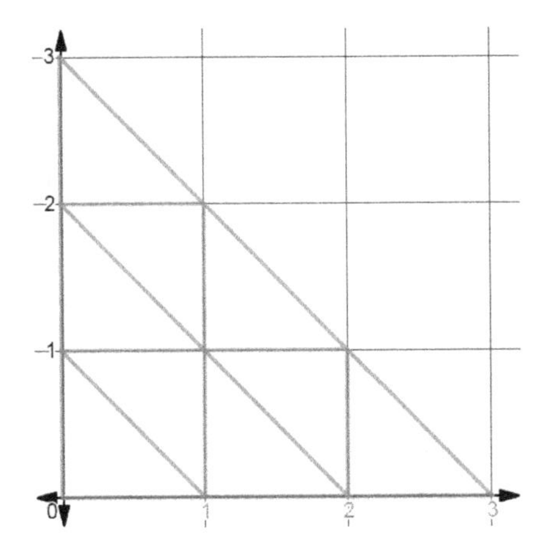

15. Write **piecewise functions** on the cognate restricted intervals that render the corresponding system of right triangles:

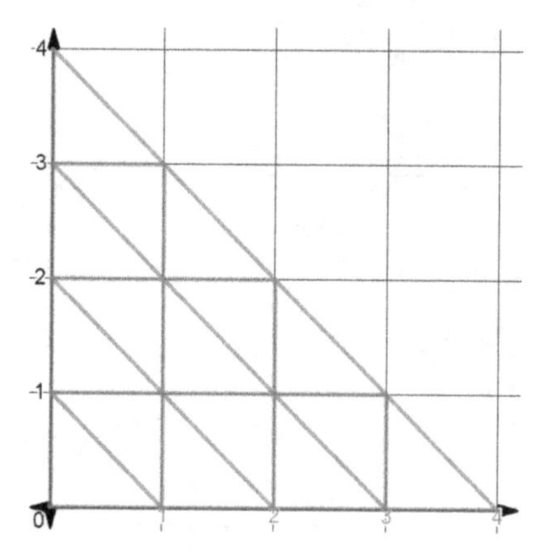

16. Write **piecewise functions** on the cognate restricted intervals that render the corresponding system of right triangles:

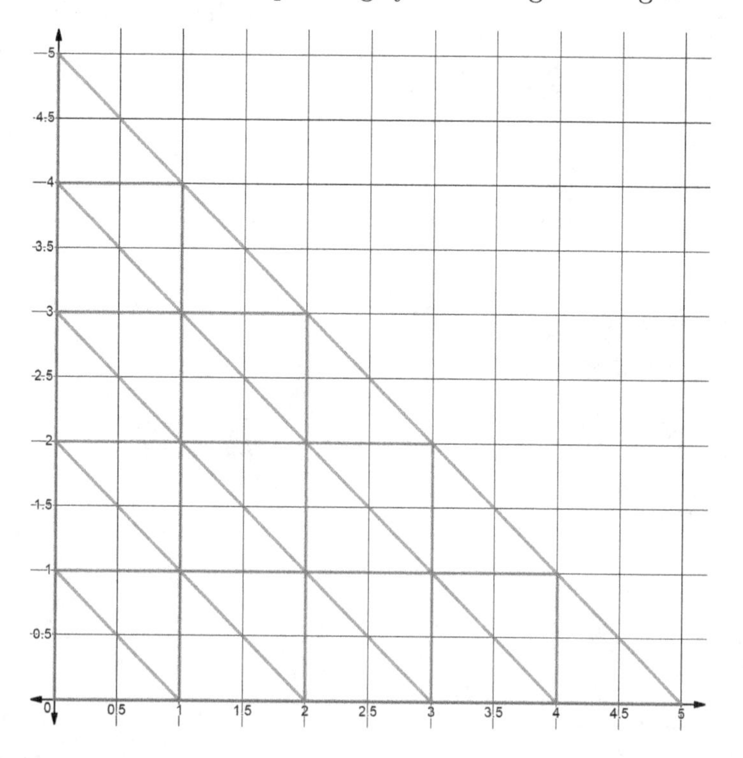

17. Write **piecewise functions** on the cognate restricted intervals that render the corresponding system of equilateral triangles:

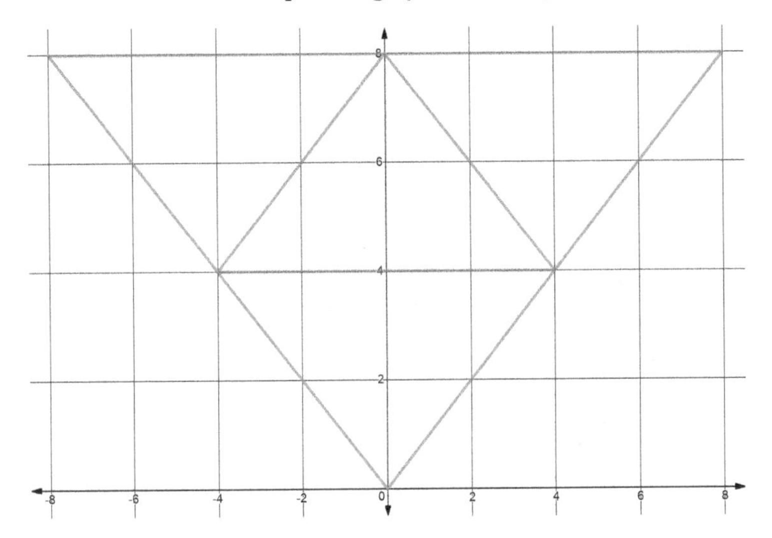

18. Write **piecewise functions** on the cognate restricted intervals that render the corresponding system of equilateral triangles:

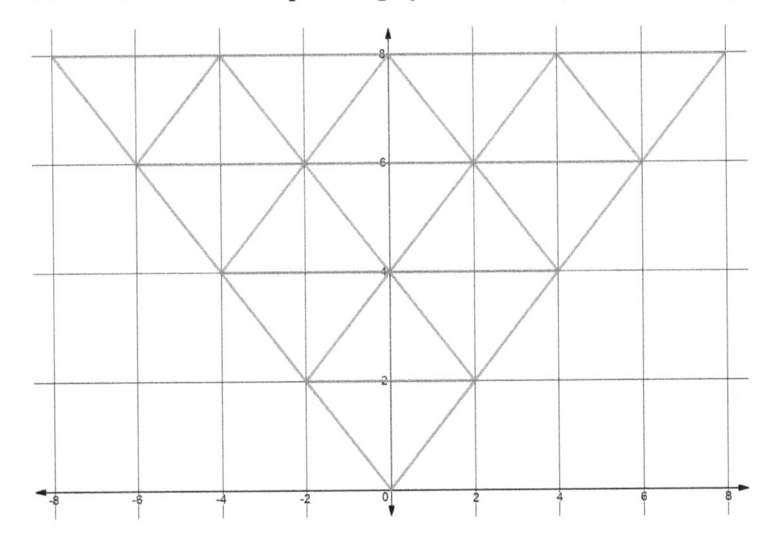

19. Write **piecewise functions** on the cognate restricted intervals that depict the corresponding system of shrinking equilateral triangles:

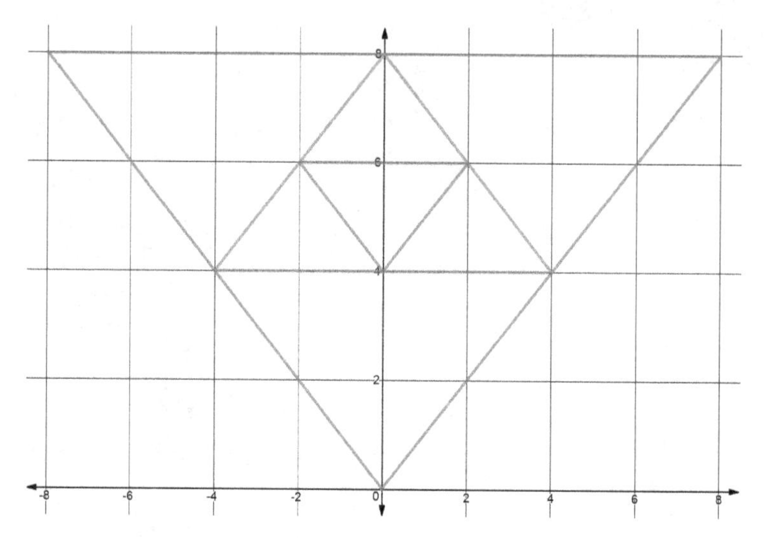

20. Write **piecewise functions** on the cognate restricted intervals that depict the corresponding system of shrinking equilateral triangles:

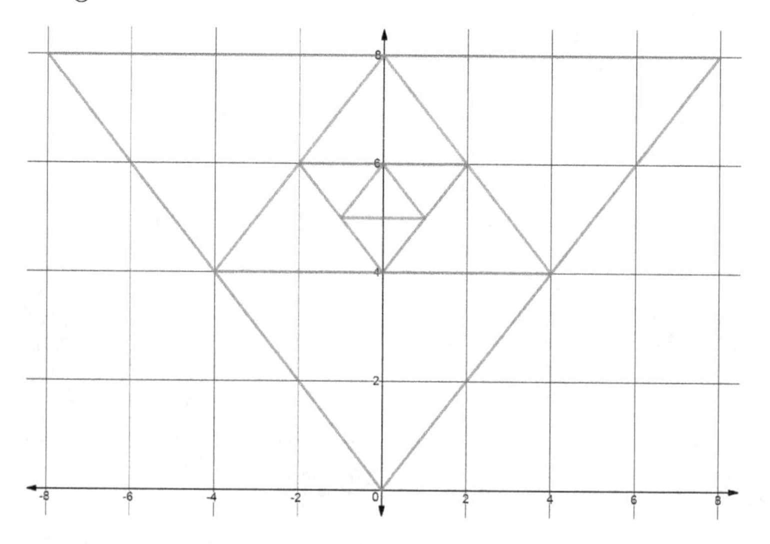

21. Write **piecewise functions** on the cognate restricted intervals that characterize the corresponding system of shrinking squares:

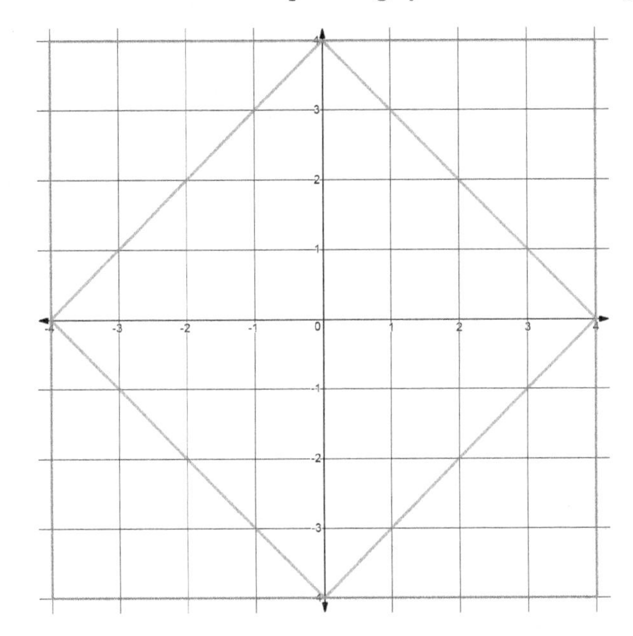

22. Write **piecewise functions** on the cognate restricted intervals that characterize the corresponding system of shrinking squares:

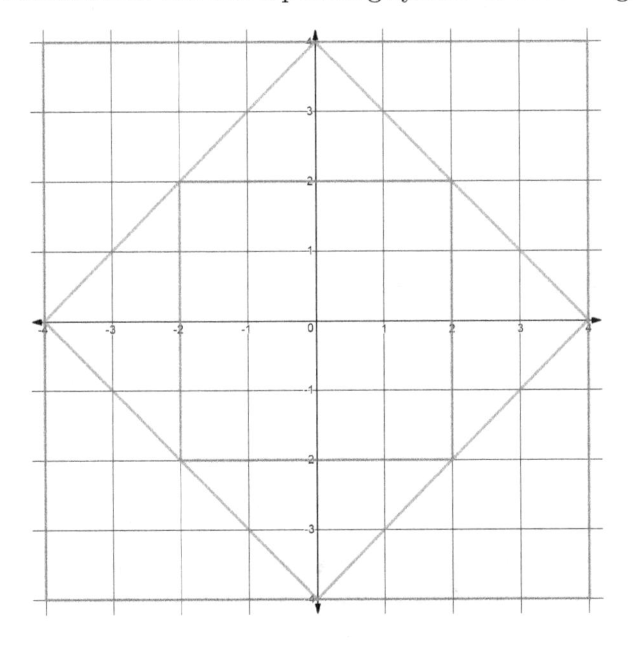

23. Write **piecewise functions** on the cognate restricted intervals that characterize the corresponding system of shrinking squares:

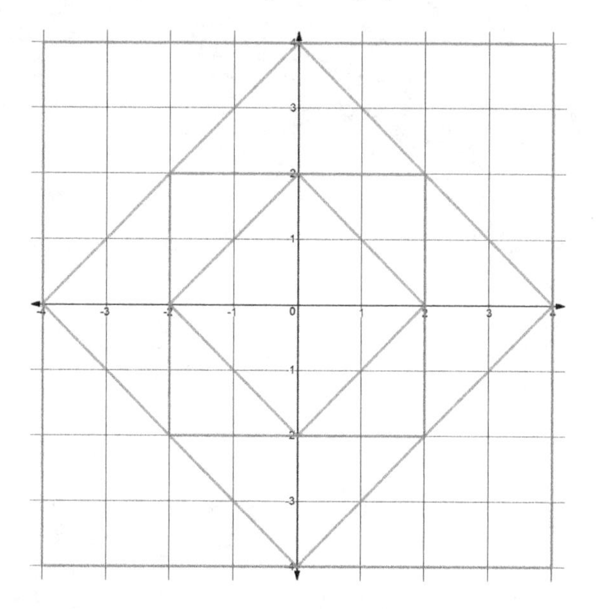

24. Write **piecewise functions** on the cognate restricted intervals that characterize the corresponding system of shrinking squares:

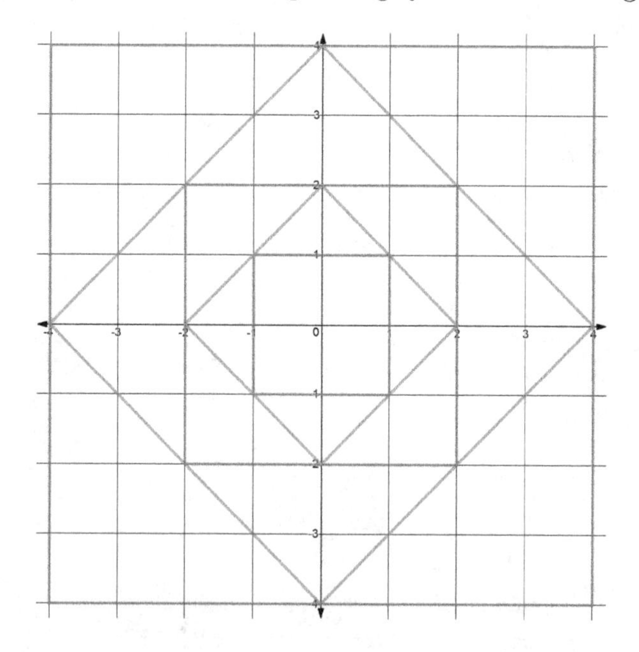

25. Write **piecewise functions** on the cognate restricted intervals that describe the corresponding system of shrinking right triangles:

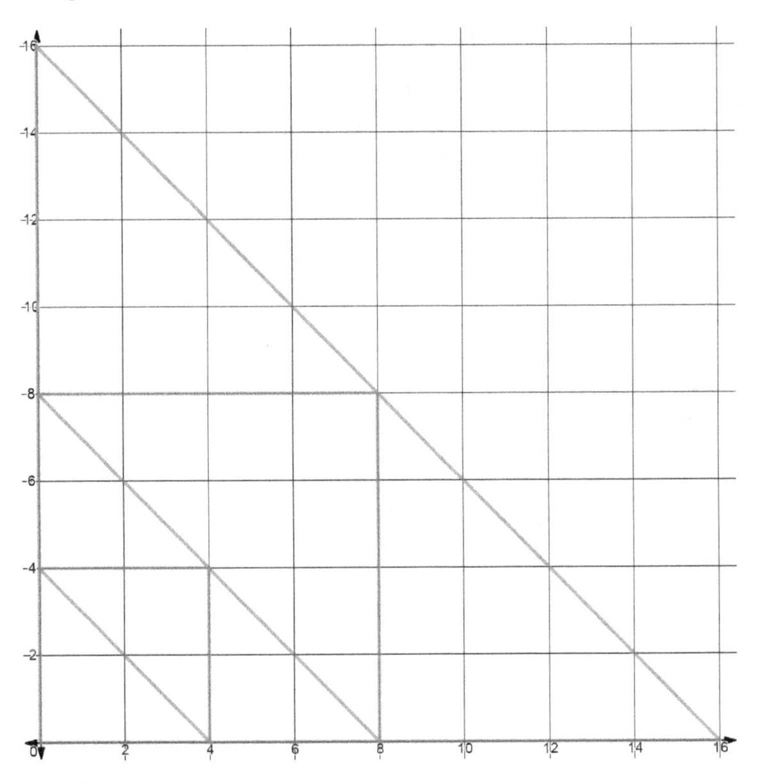

26. Write **piecewise functions** on the cognate restricted intervals that describe the corresponding system of shrinking right triangles:

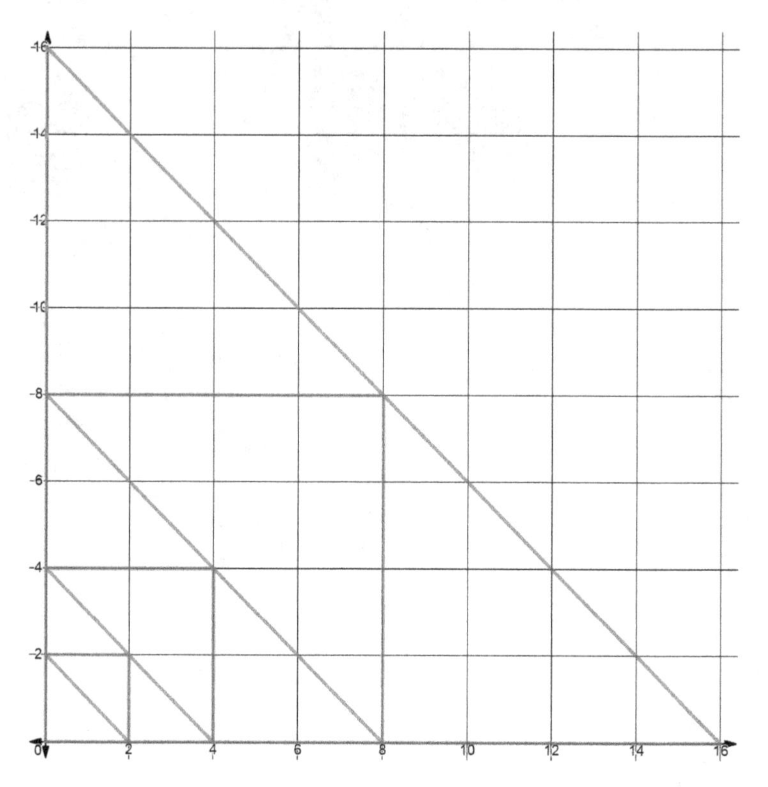

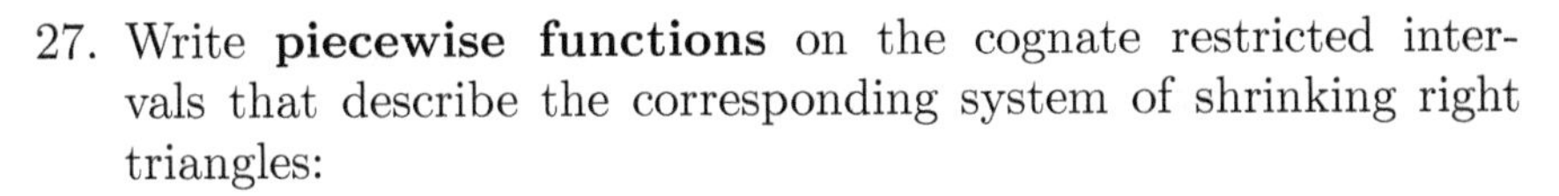

27. Write **piecewise functions** on the cognate restricted intervals that describe the corresponding system of shrinking right triangles:

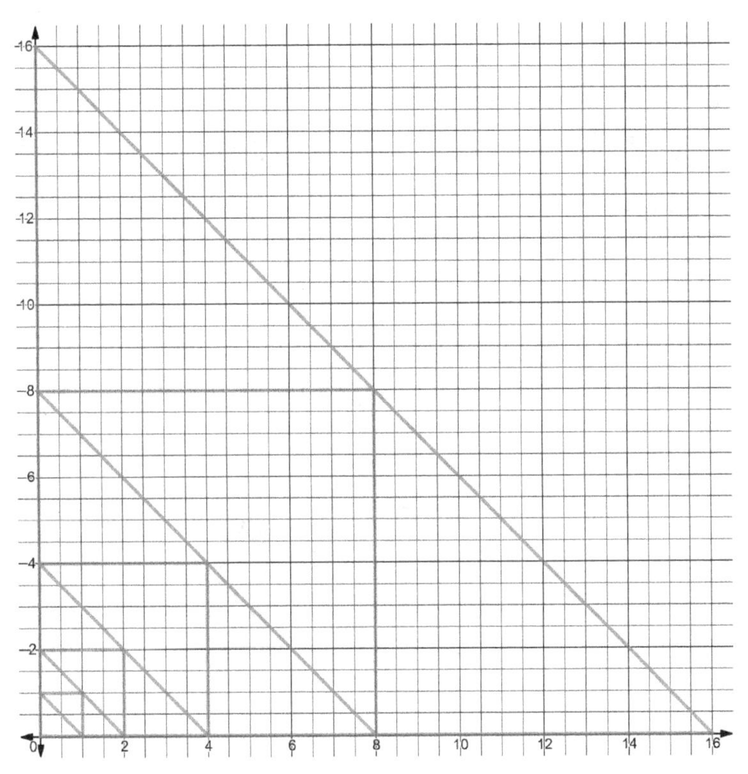

28. Write **piecewise functions** on the cognate restricted intervals that describe the corresponding system of shrinking right triangles:

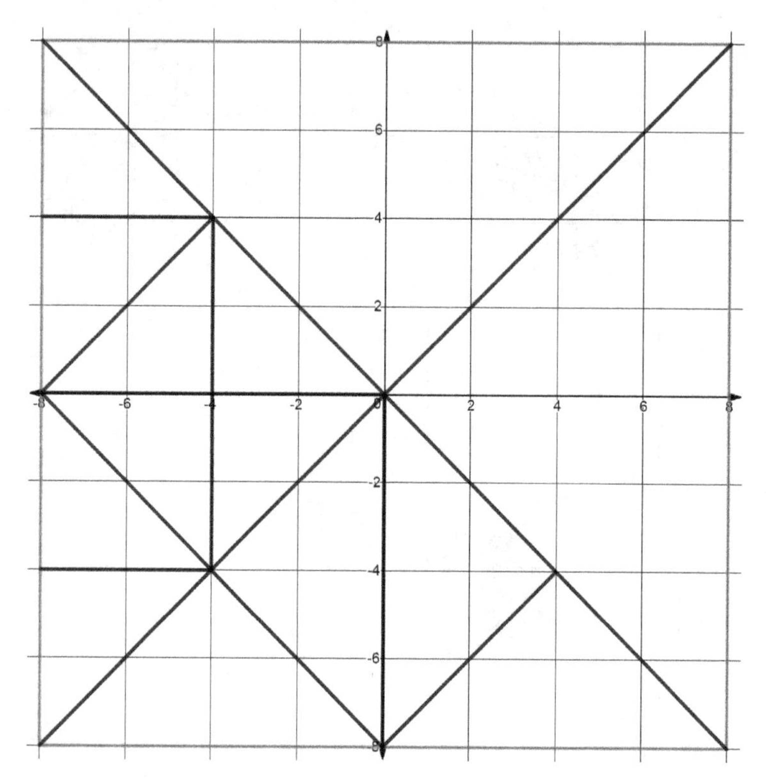

29. Write **piecewise functions** on the cognate restricted intervals that describe the corresponding system of shrinking right triangles:

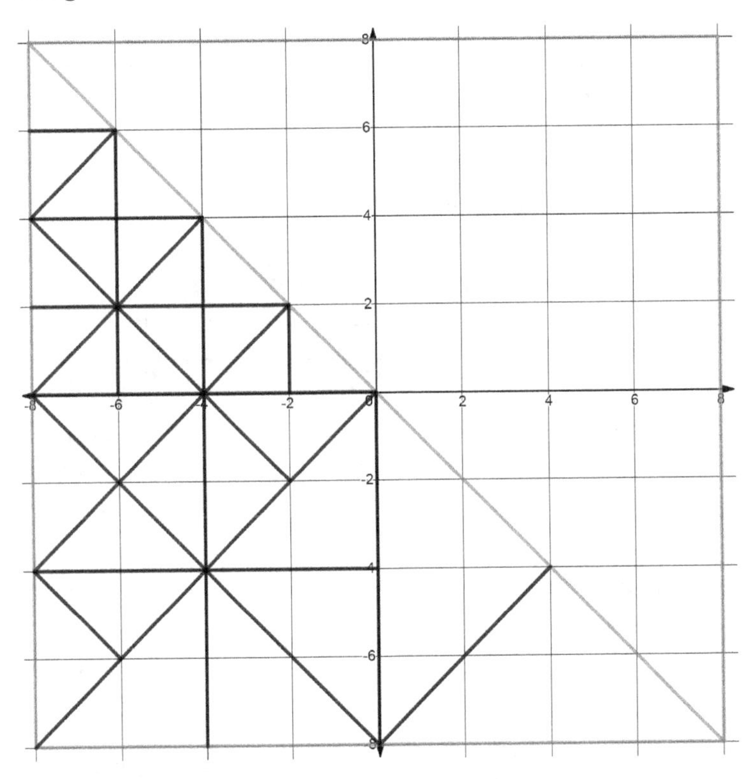

30. Write **piecewise functions** on the cognate restricted intervals that describe the corresponding system of shrinking right triangles:

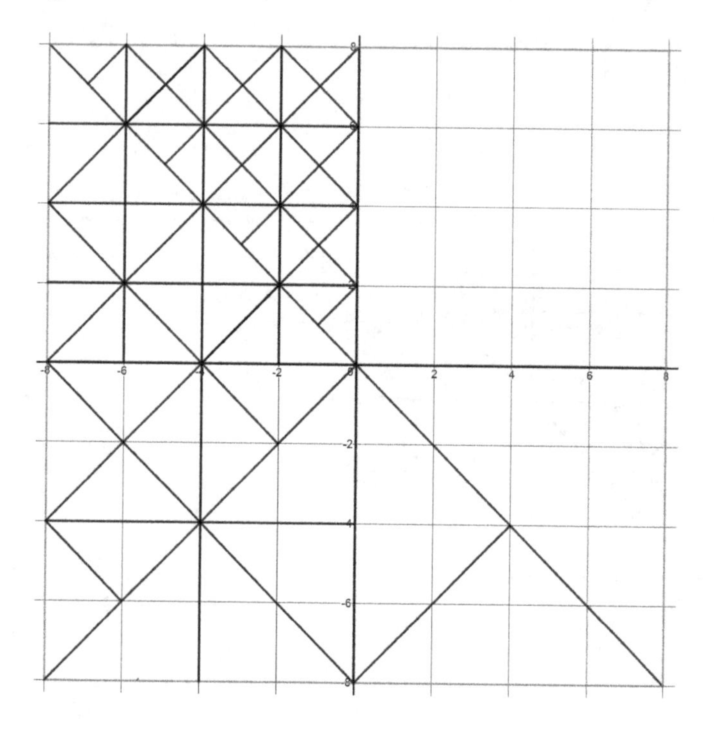

31. Write a **formula** that determines the number of inscribed blue triangles inside the main triangle:

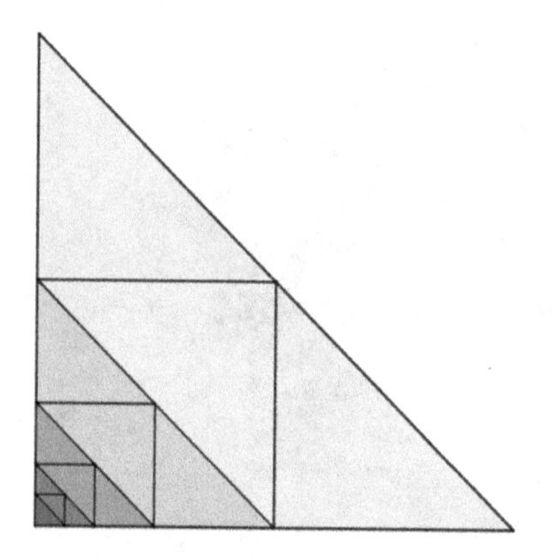

32. Write a **formula** that determines the number of inscribed blue triangles inside the main triangle:

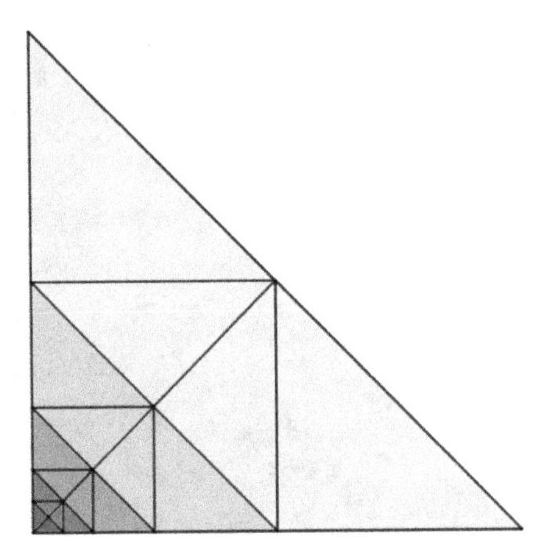

33. Solve for k when the dimension of each small square is k and the perimeter P (the sum of all the red sides) is equal to the area A of all the squares:

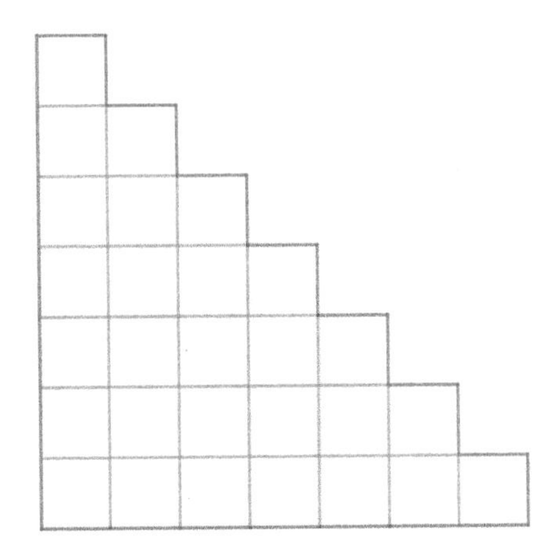

34. Using Exercise 33, suppose that there are n rows of squares ($n \geq 2$). Solve for k when the dimension of each small square is k and the perimeter P (the sum of all the red sides) is equal to the area A of all the squares.

Chapter 3

Sequences, Products and Summations

You encountered formulation of various sequences and patterns such as linear, quadratic, geometric, product-type and summation-type sequences. This chapter's intents are to decipher and formulate analogous sequences and patterns in more thorough details and to examine supplemental sequences and patterns. For example, Figure 3.1 enumerates the number of decomposed 45-45-90 triangles in between the green and blue squares.

Figure 3.1: System of diminishing squares and triangles.

Then for all $n \in \mathbb{N}$, the number of triangles in Figure 3.1 is resembled by the corresponding **linear sequence**:

$$4,\ 8,\ 12,\ 16,\ 20,\ \ldots, 4n\ =\ \{4i\}_{i=1}^{n}. \tag{3.1}$$

Observe that (3.1) enumerates the number of triangles and recites multiples of 4 starting with 4.

Figure 3.2 numerates the number of diminishing blue 45-45-90 triangles:

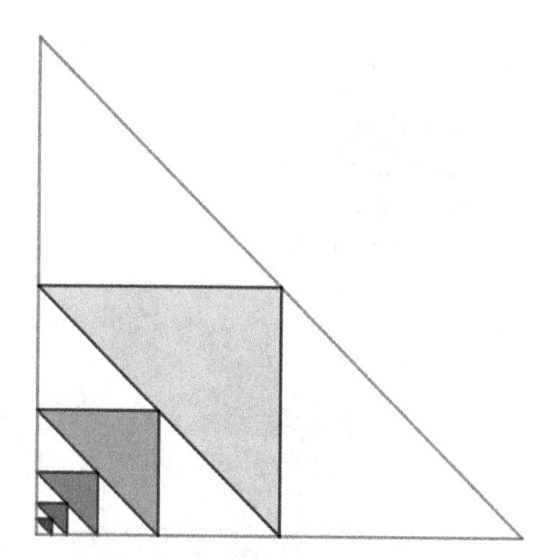

Figure 3.2: System of diminishing blue 45-45-90 triangles.

Then for all $n \in \mathbb{N}$, the areas of triangles in Figure 3.2 are described by the following **geometric sequence**:

$$2,\ \frac{1}{2},\ \frac{1}{8},\ \frac{1}{32},\ \frac{1}{128},\ \ldots,\ 2^{1-2n},\ \ldots. \tag{3.2}$$

Note that (3.2) enumerates the areas of the diminishing blue 45-45-90 triangles starting with the largest principle red 45-45-90 triangle whose area is 2.

Figure 3.3 specifies a pyramid-shaped arrangement of squares with one square in the first row, two squares in the second row, three squares in the third row, etc.

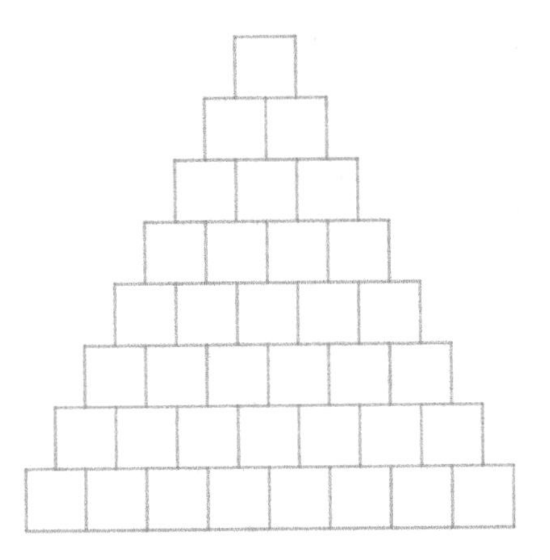

Figure 3.3: Pyramid-shaped system of squares.

Then for all $n \in \mathbb{N}$, the number of squares in Figure 3.3 is determined by the following **linear summation**:

$$1 + 2 + 3 + 4 + 5 + 6 + \cdots + n = \sum_{i=1}^{n} i = \frac{n[n+1]}{2}. \quad (3.3)$$

Note that (3.3) adds consecutive positive integers starting with 1.

Figure 3.4 determines the total number of equilateral triangles at the same scale with an arrangement into rows with an odd number of triangles in each row. One triangle in the first row, three triangles in the second row, five triangles in the third row, etc.

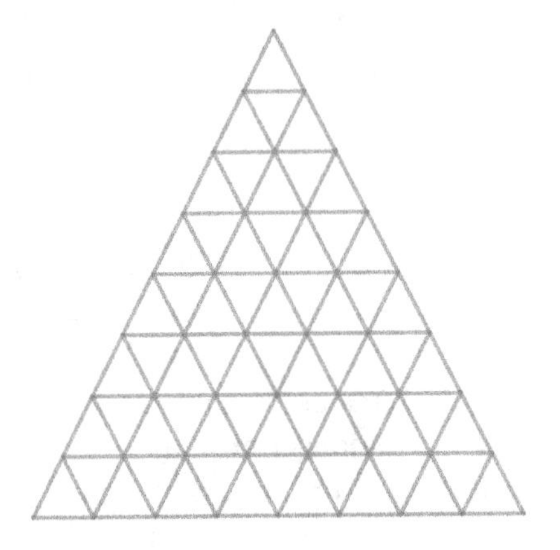

Figure 3.4: Pyramid-shaped system of triangles.

Then for all $n \in \mathbb{N}$, the number of triangles in Figure 3.4 is characterized by the corresponding **linear summation:**

$$1+3+5+7+9+11+\cdots+(2n-1) = \sum_{i=1}^{n}(2i-1) = n^2\,. \quad (3.4)$$

Note that (3.4) adds consecutive positive odd integers starting with 1, which adds up to a perfect square.

Figure 3.5 decomposes a square into a system of diminishing 45-45-90 triangles.

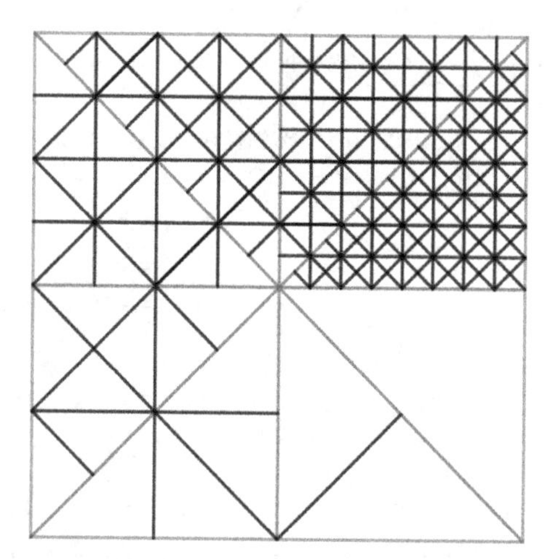

Figure 3.5: Pyramid-shaped system of triangles.

Then for all $n \in \mathbb{N}$, the number of triangles in Figure 3.5 is determined by the corresponding **geometric summation:**

$$1 + 2^1 + 2^2 + 2^3 + 2^4 + 2^5 + 2^6 + 2^7 = \sum_{i=0}^{7} 2^i = 2^8 - 1. \quad (3.5)$$

Note that 3.5 adds eight terms as the blue square is decomposed into eight primary triangular regions (emphasized by the red diagonal, horizontal and vertical lines) and into 8 groups of 45-45-90 triangles with each group at a specific scale. Then for all $n \in \mathbb{N}$ (3.5) extends to the corresponding **geometric summation**:

$$a \cdot r^0 + a \cdot r^1 + a \cdot r^2 + a \cdot r^3 + \cdots + a \cdot r^n = \sum_{i=0}^{n} a \cdot r^i$$

$$= \frac{a[1 - r^{n+1}]}{1 - r}. \tag{3.6}$$

In Chapter 4, you will apply the **proof by induction** method to prove (3.3), (3.4), and (3.6) and derive and prove supplemental summations. The upcoming Section 3.1 will investigate the formulation of linear sequences.

3.1 Linear Sequences

A **linear sequence** is constructed by either reciting multiples of a concrete constant or by adding a constant from neighbor to neighbor. You encountered examples of linear sequences in Chapters 1 and 2 analogous to the linear sequence in (3.1). The aim is to write a formula that recites all the specified terms of a given sequence with a starting term and an ending term and the analogous starting and ending indices. You will examine examples with a range of patterns. The succeeding example examines a finite linear sequence that recites multiples of 6.

Example 3.1. Write a formula of the following sequence:

$$12,\ 18,\ 24,\ 30,\ 36,\ \ldots,\ 126. \tag{3.7}$$

Solution: First, note that (3.7) commences with 12 and terminates with 126 while shifting from neighbor to neighbor by adding a 6. Next, observe that (3.7) lists values that are multiples of 6 starting with 12. Then reformulate (3.7) as:

$$6 \cdot 2,\ 6 \cdot 3,\ 6 \cdot 4,\ 6 \cdot 5,\ 6 \cdot 6,\ \ldots,\ 6 \cdot 21. \tag{3.8}$$

Hence, via (3.8) you obtain the associated formula describing (3.7):

$$\{6 \cdot i\}_{i=2}^{21}, \tag{3.9}$$

where $i = 2$ is the starting index and $i = 21$ is the terminating index. By downshifting the starting and the terminating indices of (3.9) by 1 you obtain:

$$\{6i\}_{i=2}^{21} = \{6(i+1)\}_{i=1}^{20} = \{6i + 6\}_{i=1}^{20}.$$

The upcoming examples examine an infinite linear sequence with a starting index only.

Example 3.2. Write a formula of the following sequence:

$$8,\ 13,\ 18,\ 23,\ 28,\ \ldots\ . \tag{3.10}$$

Solution: First, (3.10) commences with 8 and shifts from neighbor to neighbor by adding a 5. Next, note that (3.10) lists values that are three more than multiples of 5. Then reformulate (3.10) as:

$$5\cdot 1+3,\ 5\cdot 2+3,\ 5\cdot 3+3,\ 5\cdot 4+3,\ 5\cdot 5+3,\ \ldots. \tag{3.11}$$

Hence, via (3.11) you obtain the associated formula:

$$\{5n+3\}_{n=1}^{\infty},$$

where $n=1$ is the starting index.

Example 3.3. Write a formula of the following sequence:

$$18,\ 26,\ 34,\ 42,\ 50,\ \ldots. \tag{3.12}$$

Solution: First, 3.7 commences with 18 and shifts from neighbor to neighbor by adding an 8. Next, note that (1.20) lists values that are two more than multiples of 8. Then reformulate (3.10) as:

$$8\cdot 2+2,\ 8\cdot 3+2,\ 8\cdot 4+2,\ 8\cdot 5+2,\ 8\cdot 6+2,\ \ldots. \tag{3.13}$$

Hence, via (3.13) you obtain the associated formula depicting (3.12):

$$\{8n+2\}_{n=2}^{\infty}, \tag{3.14}$$

where $n=2$ is the starting index. Alternatively, by downshifting the starting index of (3.14) by 1 you obtain:

$$\{8n+2\}_{i=2}^{\infty} = \{8(n+1)+2\}_{n=1}^{\infty} = \{8n+10\}_{n=1}^{\infty}.$$

The next section will focus on the formulation of quadratic sequences.

3.2 Quadratic Sequences

A **quadratic sequence** is assembled by either listing perfect squares such as (3.4), odd perfect squares or even perfect squares. The first example resembles a sequence listing the consecutive perfect squares between 16 and 2,500.

Example 3.4. Write a formula of the following sequence:

$$16,\ 25,\ 36,\ 49,\ 64,\ 81,\ \ldots,\ 2{,}500. \tag{3.15}$$

Solution: Equation (3.15) commences with 16 and terminates with 2,500 and (3.15) recites all the perfect squares between 16 and 2,500. Then reformulate (3.15) as:

$$(4)^2,\ (5)^2,\ (6)^2,\ (7)^2,\ (8)^2,\ (9)^2,\ \ldots,\ (50)^2. \tag{3.16}$$

Hence, via (3.16) you obtain the associated formula:

$$\{(i)^2\}_{i=4}^{50}, \tag{3.17}$$

where $i = 4$ is the starting index and $i = 50$ is the terminating index. By downshifting the starting index of (3.17) by 3 you obtain:

$$\{(i)^2\}_{i=4}^{50} = \{(i+3)^2\}_{i=1}^{47}.$$

The succeeding example examines an infinite quadratic sequence describing the consecutive even perfect squares starting with 4.

Example 3.5. Write a formula of the following sequence:

$$4,\ 16,\ 36,\ 64,\ 100,\ 144,\ \ldots. \tag{3.18}$$

Solution: Equation (3.18) recites all the even perfect squares starting with 4. First, reformulate (3.18) as:

$$(2)^2,\ (4)^2,\ (6)^2,\ (8)^2,\ (10)^2,\ (12)^2,\ \ldots. \tag{3.19}$$

Next, reformulate (3.19) as:

$$(2\cdot 1)^2, (2\cdot 2)^2, (2\cdot 3)^2, (2\cdot 4)^2, (2\cdot 5)^2, (2\cdot 6)^2, \ldots. \tag{3.20}$$

Hence, via (3.20) you obtain the associated formula:

$$\{(2\cdot n)^2\}_{n=1}^{\infty},$$

where $n = 1$ is the starting index.

The upcoming example analyzes an infinite quadratic sequence describing the consecutive odd perfect squares starting with 1.

Example 3.6. Write a formula of the following sequence:

$$1,\ 9,\ 25,\ 49,\ 81,\ 121,\ \ldots. \tag{3.21}$$

Solution: Equation (3.21) recites all the odd perfect squares starting with 1. First reformulate (3.21) as:

$$(1)^2,\ (3)^2,\ (5)^2,\ (7)^2,\ (9)^2,\ (11)^2,\ \ldots. \tag{3.22}$$

Next, reformulate (3.22) as:

$$(2\cdot 1-1)^2,\ (2\cdot 2-1)^2,\ (2\cdot 3-1)^2,\ (2\cdot 4-1)^2,\ (2\cdot 5-1)^2,\ (2\cdot 6-1)^2,\ldots. \tag{3.23}$$

Hence, via (3.23) you obtain the associated formula:

$$\{(2n-1)^2\}_{n=1}^{\infty},$$

where $n = 1$ is the starting index.

The sequence $\{n^2\}_{n=1}^{\infty}$ recites the consecutive perfect squares and guides you to the upcoming Section 3.3 as it is a special case of the **summation-type** sequences that we encountered in (3.4).

3.3 Summation-Type Sequences

The consequent two examples formulate a **summation-type** sequence analogous to (3.3) and (3.4). The next two examples will focus on simplifying a summation-type sequence (1.41) in the form:

$$S = a_1 + a_2 + a_3 + \cdots + a_n = \sum_{i=1}^{n} a_i. \tag{3.24}$$

Note that the left-hand side of Eq. (3.24) is in the **expanded form**, while the right-hand side of Eq. (3.24) is in the **factored form**. The consequent example adds the consecutive positive even integers starting with 2 (adding a 4 to the first neighbor, adding a 6 to the next neighbor, etc.).

Example 3.7. Write a formula of the following sequence:

$$2,\ 6,\ 12,\ 20,\ 30,\ 42,\ 56,\ \ldots. \tag{3.25}$$

Solution: Note that (3.25) commences with 2, you then add a 4 to transition to the next neighbor, add a 6 to the consequent neighbor and so on. Denote $n \in \mathbb{N}$ as the terminating index and formulate (3.25) as:

$$\begin{aligned}
&2,\\
&2+4=6,\\
&6+6=[\mathbf{2}+\mathbf{4}]+6=12,\\
&12+8=[\mathbf{2}+\mathbf{4}+\mathbf{6}]+8=20,\\
&20+10=[\mathbf{2}+\mathbf{4}+\mathbf{6}+\mathbf{8}]+10=30,\\
&30+12=[\mathbf{2}+\mathbf{4}+\mathbf{6}+\mathbf{8}+\mathbf{10}]+12=42,\\
&42+14=[\mathbf{2}+\mathbf{4}+\mathbf{6}+\mathbf{8}+\mathbf{10}+\mathbf{12}]+14=56,\\
&\vdots
\end{aligned} \tag{3.26}$$

Thus, via (3.26), for $n \in \mathbb{N}$ you obtain the associated formula:

$$\{a_i\}_{i=1}^{n} = \sum_{i=1}^{n} 2i.$$

In Chapter 6, you will reformulate (3.25) as a recursive sequence.

The consequent example will require the use of summations and a piecewise sequence.

Example 3.8. Write a formula of the following sequence:

$$1,\ 2,\ 5,\ 10,\ 17,\ 26,\ 37,\ \ldots. \tag{3.27}$$

Solution: Note that (3.27) commences with 1, you then add a 1 to transition to the next neighbor, add a 3 to the consequent neighbor and so on. Then formulate (3.27) as:

$$\begin{aligned}
&1,\\
&1+[\mathbf{1}]=2,\\
&2+3=1+[\mathbf{1}+\mathbf{3}]=5,\\
&5+5=1+[\mathbf{1}+\mathbf{3}+\mathbf{5}]=10,\\
&10+7=1+[\mathbf{1}+\mathbf{3}+\mathbf{5}+\mathbf{7}]=17,\\
&17+9=1+[\mathbf{1}+\mathbf{3}+\mathbf{5}+\mathbf{7}+\mathbf{9}]=26,\\
&26+11=1+[\mathbf{1}+\mathbf{3}+\mathbf{5}+\mathbf{7}+\mathbf{9}+\mathbf{11}]=37,\\
&\vdots
\end{aligned} \tag{3.28}$$

Thus, via (3.28), for $n \geq 0$ you acquire the associated **piecewise formula:**

$$\{a_i\}_{i=0}^{n} = \begin{cases} 1 & \text{if } n = 0, \\ 1 + \sum_{i=1}^{n} (2i-1) & \text{if } n \in \mathbb{N}. \end{cases}$$

You will seek additional examples of piecewise sequences in latter part of this chapter and in latter chapters. In Chapter 6, you will render piecewise sequences as recursive sequences.

The upcoming Section 3.4 will focus on the characteristics of **geometric** sequences.

3.4 Geometric Sequences

A **geometric sequence** is contrived by multiplying by a constant from term to term as you observed in various examples such as (3.2). For $n \geq 0$, we define an **infinite geometric sequence** in the corresponding form:

$$\{a \cdot r^n\}_{n=0}^{\infty} = a \cdot r^0,\ a \cdot r^1,\ a \cdot r^2,\ a \cdot r^3,\ a \cdot r^4,\ \ldots, \tag{3.29}$$

where a is the **starting term** of (3.29) and r is the **multiplicative factor** of (3.29), where $r \neq 1$. The upcoming example examines a geometric sequence by multiplying from neighbor to neighbor by 2.

Example 3.9. Write a formula of the following sequence:

$$15,\ 30,\ 60,\ 120,\ 240,\ \ldots\ . \tag{3.30}$$

Solution: (3.30) is a geometric sequence with the starting term $a = 15$ and the multiplicative factor $r = 2$. Thus, you transition from neighbor to neighbor by multiplying by 2 and reformulate (3.30) as:

$$15,\ 15 \cdot 2,\ 30 \cdot 2,\ 60 \cdot 2,\ 120 \cdot 2,\ \ldots. \tag{3.31}$$

Next, reformulate (3.31) as:

$$15 \cdot 2^0,\ 15 \cdot 2^1,\ 15 \cdot 2^2,\ 15 \cdot 2^3,\ 15 \cdot 2^4,\ \ldots. \tag{3.32}$$

For all $n \geq 0$ formulate (3.32) as:

$$\{15 \cdot 2^n\}_{n=0}^{\infty}. \tag{3.33}$$

By rewriting the index by 1, you reformulate (3.33) as:

$$\{15 \cdot 2^{n-1}\}_{n=1}^{\infty}\ .$$

The next example will examine and formulate a geometric sequence with square root algebra.

Example 3.10. Write a formula of the following sequence:

$$3,\ 3\sqrt{3},\ 9,\ 9\sqrt{3},\ 27,\ \ldots. \tag{3.34}$$

Solution: Equation (3.34) is a geometric sequence with the starting term $a = 3$ and the multiplicative factor $r = \sqrt{3}$. Thus, you transition from neighbor to neighbor by multiplying by $\sqrt{3}$ and reformulate (3.34) as:

$$3,\ 3 \cdot \sqrt{3},\ 3\sqrt{3} \cdot \sqrt{3},\ 9 \cdot \sqrt{3},\ 9\sqrt{3} \cdot \sqrt{3},\ \ldots. \tag{3.35}$$

Next, reformulate (3.35) as:

$$3 \cdot \left[\sqrt{3}\right]^0,\ 3 \cdot \left[\sqrt{3}\right]^1,\ 3 \cdot \left[\sqrt{3}\right]^2,\ 3 \cdot \left[\sqrt{3}\right]^3,\ 3 \cdot \left[\sqrt{3}\right]^4,\ \ldots. \tag{3.36}$$

For all $n \geq 0$, formulate (3.36) as:

$$\left\{3 \cdot \left[\sqrt{3}\right]^n\right\}_{n=0}^{\infty}. \tag{3.37}$$

Alternatively you can reformulate (3.37) as:

$$\left\{\left[\sqrt{3}\right]^n\right\}_{n=2}^{\infty}\ .$$

Geometric sequences are a special case of the **product-type** sequences and hence direct you to the next section.

3.5 Product-Type Sequences

This section's aims are to investigate the characteristics and formulations of **product-type sequences**. These include the **factorial pattern** and the **factorial-type** patterns. The next two examples will focus on simplifying a product-type sequence (1.41) in the form:

$$P = a_1 \cdot a_2 \cdot a_3 \cdot \cdots \cdot a_n = \prod_{i=1}^{n} a_i. \tag{3.38}$$

Note that the left-hand side of Eq. (3.38) is in the **expanded form** while the right-hand side of Eq. (3.38) is in the **factored form**.

Example 3.11. In terms of $n \in \mathbb{N}$, write a formula of the following sequence:

$$\prod_{i=1}^{n} \left[\frac{i}{i+1}\right]. \tag{3.39}$$

Solution: The expanded form of (3.39) gives you:

$$\prod_{i=1}^{n} \left[\frac{i}{i+1}\right] = \frac{1}{2} \cdot \frac{2}{3} \cdot \frac{3}{4} \cdot \frac{4}{5} \cdot \frac{5}{6} \cdots \frac{n-1}{n} \cdot \frac{n}{n+1}. \tag{3.40}$$

Next, observe that (3.40) simplifies to:

$$\frac{1}{n+1}.$$

The succeeding example will apply the sum of consecutive integers (3.3) to formulate a product-type sequence.

Example 3.12. In terms of $n \in \mathbb{N}$, write a formula of the following sequence:

$$\prod_{i=1}^{n} 2^i. \tag{3.41}$$

Solution: The expanded form of (3.41) gives you:

$$\prod_{i=1}^{n} 2^i = 2^1 \cdot 2^2 \cdot 2^3 \cdot 2^4 \cdot 2^5 \cdots 2^{n-1} \cdot 2^n. \tag{3.42}$$

Using laws of exponents you reformulate (3.42) as:

$$2^{(1\ +\ 2\ +\ 3\ +\ 4\ +\ \cdots\ +\ n)}. \tag{3.43}$$

Next, using (3.3) and laws of exponents, (3.43) reduces to the corresponding expression:

$$2^{\frac{n\cdot(n+1)}{2}} = \left(\sqrt{2}\right)^{n\cdot(n+1)}.$$

The next example will investigate a special case of product-type sequence in terms of factorials.

Example 3.13. Rewrite the following sequence as a product-type sequence of factorials:

$$\prod_{i=1}^{n} i^i. \tag{3.44}$$

Solution: First, express the expanded form of (3.44) in the corresponding **lower-triangular matrix form**:

$$\begin{array}{l}
1 \\
2 \cdot 2 \\
3 \cdot 3 \cdot 3 \\
4 \cdot 4 \cdot 4 \cdot 4 \\
5 \cdot 5 \cdot 5 \cdot 5 \cdot 5 \\
6 \cdot 6 \cdot 6 \cdot 6 \cdot 6 \cdot 6 \\
7 \cdot 7 \cdot 7 \cdot 7 \cdot 7 \cdot 7 \cdot 7 \\
\vdots \quad \vdots \quad \vdots \quad \vdots \quad \vdots \quad \vdots \quad \vdots
\end{array} \tag{3.45}$$

Next, depict (3.45) as the cognate system of products that multiply all the consecutive integers in each column in (3.45). Starting with 1 in the first column, starting with 2 in the second column, starting with 3 in the third column, and so on. Hence, reformulate (3.45) as:

$$\prod_{i=1}^{n} i \cdot \prod_{i=2}^{n} i \cdot \prod_{i=3}^{n} i \cdot \prod_{i=4}^{n} i \cdot \cdots \cdot \prod_{i=n}^{n} i. \tag{3.46}$$

Using the attributes of factorials in Example (1.9), revise (3.46) as a product of corresponding factorials:

$$\frac{n!}{0!} \cdot \frac{n!}{1!} \cdot \frac{n!}{2!} \cdot \frac{n!}{3!} \cdot \cdots \cdot \frac{n!}{(n-1)!} \tag{3.47}$$

which then simplifies to the associated product-type and factorial-type sequence:

$$\prod_{i=1}^{n} \frac{n!}{(i-1)!}.$$

Example 3.13 then directs you to factorial-type sequences.

3.5.1 *Factorial-type Sequences*

This section's aims are to examine the traits of **factorial-type** sequences, which are special cases of the product-type sequences. In Chapter 1, for all $n \in \mathbb{N}$ you contrived the **factorial pattern** by multiplying the consecutive positive integers starting at 1 as:

$$0! = 1 \quad \text{and} \quad n! = \prod_{i=1}^{n} i. \tag{3.48}$$

Alternatively, for all $n \geq 0$, (3.48) can be formulated as the corresponding **piecewise sequence**:

$$n! = \begin{cases} 1 & \text{if } n = 0, \\ \prod_{i=1}^{n} i & \text{if } n \in \mathbb{N}. \end{cases} \tag{3.49}$$

Note that (3.49) is similar to (3.3). In fact, (3.3) adds consecutive positive integers, while (3.49) multiplies consecutive positive integers. Now we will transition to assorted examples of patterns that mimic the **factorial pattern**. In Chapter 6, you will reformulate the Factorial as a recursive sequence by multiplying terms instead of adding terms. The upcoming example will analyze the product of consecutive positive odd integers. Analogous to Example 3.8, the upcoming example examines the geometric-type pattern that guides you to the **factorial-type** sequences.

Example 3.14. Write a formula of the following sequence:

$$1,\ 3,\ 15,\ 105,\ 945,\ 10{,}395,\ \ldots. \tag{3.50}$$

Solution: Note that (3.50) commences with 1, you then transition to the next neighbor by multiplying by 3, multiply the consequent neighbor by 5 and so on. Denote $n \in \mathbb{N}$ as the terminating index and formulate (3.50) as:

$$\begin{aligned}
&1,\\
&[\mathbf{1} \cdot \mathbf{3}] = 3,\\
&3 \cdot 5 = [\mathbf{1} \cdot \mathbf{3} \cdot \mathbf{5}] = 15,\\
&15 \cdot 7 = [\mathbf{1} \cdot \mathbf{3} \cdot \mathbf{5} \cdot \mathbf{7}] = 105,\\
&105 \cdot 9 = [\mathbf{1} \cdot \mathbf{3} \cdot \mathbf{5} \cdot \mathbf{7} \cdot \mathbf{9}] = 945,\\
&945 \cdot 11 = [\mathbf{1} \cdot \mathbf{3} \cdot \mathbf{5} \cdot \mathbf{7} \cdot \mathbf{9} \cdot \mathbf{11}] = 10{,}395,\\
&\vdots
\end{aligned} \tag{3.51}$$

Observe that (3.51) is analogous to (3.4). In fact, 3.4 adds consecutive positive odd integers, while (3.51) multiplies consecutive positive odd integers. Thus via (3.51), for $n \in \mathbb{N}$ you obtain the associated formula:

$$\prod_{i=1}^{n} (2i - 1).$$

In the succeeding example you will apply a **piecewise sequence** to formulate the assigned sequence.

Example 3.15. Write a formula of the following sequence:

$$2,\ 10,\ 70,\ 630,\ 6{,}930,\ \ldots. \tag{3.52}$$

Solution: Observe that (3.52) commences with 2 and then shifts to the next neighbor by multiplying by 5, then multiplying by 7 to the consequent neighbor and so on. Denote $n \in \mathbb{N}$ as the terminating index and formulate (3.52) as:

$$\begin{aligned}
&2,\\
&2 \cdot [\mathbf{5}] = 10,\\
&10 \cdot 7 = 2 \cdot [\mathbf{5} \cdot \mathbf{7}] = 70,\\
&70 \cdot 9 = 2 \cdot [\mathbf{5} \cdot \mathbf{7} \cdot \mathbf{9}] = 630,\\
&630 \cdot 11 = 2 \cdot [\mathbf{5} \cdot \mathbf{7} \cdot \mathbf{9} \cdot \mathbf{11}] = 6{,}930,\\
&\vdots
\end{aligned} \tag{3.53}$$

Thus, via (3.53), for $n \in \mathbb{N}$ you obtain the associated formula depicting (3.52):

$$\{x_n\}_{i=1}^{n} = \begin{cases} 2 & \text{if } n = 1, \\ 2 \cdot \left[\prod_{i=2}^{n} (2i+1)\right] & \text{if } n \geq 2. \end{cases} \tag{3.54}$$

Observe that the product formula in (3.54) works only starting with the second term of the sequence and therefore requires the formulation of a piecewise formula.

Examples (3.8) and (3.15) are special cases of **piecewise sequences** which then lead you to the upcoming section.

3.6 Alternating and Piecewise Sequences

This section will focus on the features and formulations of **piecewise functions**. In Chapter 1, you came upon the following **alternating piecewise sequence** that diverts between 1 and -1, where for all $n \geq 0$:

$$\{(-1)^n\}_{n=0}^{\infty} = \begin{cases} 1 & \text{if } n = 0, 2, 4, 6, \ldots, \\ -1 & \text{if } n = 1, 3, 5, 7, \ldots. \end{cases}$$

Furthermore, you encountered an alternating geometric sequences in Example (2.2). Alternating sequences can be written in one fragment as the sequence above. However, several piecewise sequences cannot be written in one fragment as you experienced in Examples (3.8) and (3.15). The upcoming example will examine and formulate an alternating geometric sequence.

Example 3.16. Write a formula of the following sequence:

$$3, \ -6, \ 12, \ -24, \ 48, \ -96, \ \ldots. \tag{3.55}$$

Solution: The first term of (3.55) is positive and the sign then reverses from neighbor to neighbor. Next, note that (3.55) is a geometric sequence with $a = 3$ and $r = -2$. Hence, for $n \geq 0$ formulate (3.55) as:

$$\{x_n\}_{n=0}^{\infty} = \{3 \cdot (-2)^n\}_{n=0}^{\infty} = \{3 \cdot 2^n \cdot (-1)^n\}_{n=0}^{\infty}. \tag{3.56}$$

Next, decompose (3.55) into the corresponding two even-indexed and odd-indexed sub-sequences:

$$3, \ -6, \ 12, \ -24, \ 48, \ -96, \ \ldots. \tag{3.57}$$

Thus, reformulate (3.57) as the cognate **piecewise geometric sequence**:

$$\{x_n\}_{n=0}^{\infty} = \begin{cases} \{3 \cdot 2^n\}_{n=0}^{\infty} & \text{if } n = 0, 2, 4, 6, \ldots, \\ \\ \{-6 \cdot 2^{n-1}\}_{n=1}^{\infty} & \text{if } n = 1, 3, 5, 7, \ldots. \end{cases}$$

Example 3.16 transitions to the corresponding **alternating geometric sequences**:

$$a, \ -a \cdot r, \ a \cdot r^2, \ -a \cdot r^3, \ a \cdot r^4, \ \ldots = \{a \cdot r^n \cdot (-1)^n\}_{n=0}^{\infty},$$

and

$$-a, \ a \cdot r, \ -a \cdot r^2, \ a \cdot r^3, \ -a \cdot r^4, \ \ldots = \{a \cdot r^n \cdot (-1)^{n+1}\}_{n=0}^{\infty}.$$

Analogous to Figure 2.11, the upcoming example will examine and formulate an alternating linear sequence.

Example 3.17. Write a formula of the following sequence:

$$-9, \ 18, \ -27, \ 36, \ -45, \ 54, \ \ldots. \tag{3.58}$$

Solution: (3.58) recites multiples of 9 starting with a negative term and then the sign reverses to positive and so on. Thus, for all $n \in \mathbb{N}$

you procure:

$$\{x_n\}_{n=1}^{\infty} = \{(-1)^n \, 9n\}_{n=1}^{\infty}. \tag{3.59}$$

Next, decompose (3.59) into the corresponding two even-indexed and odd-indexed sub-sequences:

$$-9, \; 18, \; -27, \; 36, \; -45, \; 54, \; \ldots. \tag{3.60}$$

Thus, reformulate (3.60) as the corresponding **piecewise linear sequence**:

$$\{x_n\}_{n=0}^{\infty} = \begin{cases} -[9(n+1)] & \text{if } n = 0, 2, 4, 6, \ldots, \\ 9(n+1) & \text{if } n = 1, 3, 5, 7, \ldots. \end{cases}$$

The consequent example will require decomposition into two sub-sequences as every fourth term of the sequence alternates in sign. In fact, you will discover decomposition into an alternating sub-sequence and a non-alternating sub-sequence.

Example 3.18. Write a formula of the following sequence:

$$2, \; 4, \; 6, \; -8, \; 10, \; 12, \; 14, \; -16, \; \ldots. \tag{3.61}$$

Solution: First, of all, (3.61) is assembled in terms of even integers starting at 2 while every fourth term of (3.61) is negative. This is a quite a contrast to what you encountered in Examples (3.16) and (3.17). Thus, decompose (3.61) into two primary blue and green sub-sequences as:

$$2, \; 4, \; 6, \; -8, \; 10, \; 12, \; 14, \; -16, \; \ldots. \tag{3.62}$$

Next, observe that the blue sub-sequence is a non-alternating sequence of (3.62), while the green sub-sequence of (3.62) alternates. Hence, for $n \geq 0$ reformulate (3.62) as the corresponding **piecewise sequence**:

$$\{x_n\}_{n=0}^{\infty} = \begin{cases} 2(n+1) & \text{if } n = 0, 2, 4, 6, \ldots, \\ (-1)^{\frac{n-1}{2}}[2(n+1)] & \text{if } n = 1, 3, 5, 7, \ldots. \end{cases}$$

3.7 Exercises

In problems 1–12, write a **formula** of each sequence:

1. $7,\ 15,\ 23,\ 31,\ 39,\ 47,\ 55,\ \ldots.$
2. $4,\ 11,\ 18,\ 25,\ 32,\ 39,\ 46,\ \ldots.$
3. $7,\ 21,\ 35,\ 49,\ 63,\ 77,\ 91,\ \ldots.$
4. $(m+8),\ (m+12),\ (m+16),\ (m+20),\ (m+24),\ \ldots.$
5. $1,\ 9,\ 25,\ 49,\ 81,\ 121,\ 169,\ \ldots.$
6. $1,\ 25,\ 81,\ 169,\ 289,\ 441,\ 625,\ \ldots.$
7. $9,\ 49,\ 121,\ 225,\ 361,\ 529,\ 729,\ \ldots.$
8. $16,\ 64,\ 144,\ 256,\ 400,\ 576,\ 784,\ \ldots.$
9. $4,\ 36,\ 100,\ 196,\ 324,\ 484,\ 676,\ \ldots.$
10. $2,\ 6,\ 12,\ 20,\ 30,\ 42,\ 56,\ \ldots.$
11. $4,\ 7,\ 12,\ 19,\ 28,\ 39,\ 52,\ \ldots.$
12. $5,\ 8,\ 14,\ 23,\ 35,\ 50,\ 68,\ \ldots.$

In problems 13–22, write a **formula** of each sequence:

13. $2,\ 6,\ 18,\ 54,\ 162,\ 486,\ 1{,}458,\ \ldots.$
14. $2,\ \sqrt{8},\ 4,\ \sqrt{32},\ 8,\ \sqrt{128},\ 16,\ \ldots.$
15. $2,\ \sqrt{12},\ 6,\ \sqrt{108},\ 18,\ \sqrt{972},\ 54,\ \ldots.$
16. $162,\ 54,\ 18,\ 6,\ 2,\ \frac{2}{3},\ \frac{2}{9},\ \ldots.$
17. $12,\ 18,\ 27,\ \frac{81}{2},\ \frac{243}{4},\ \frac{729}{8},\ \frac{2187}{16},\ \ldots.$
18. $45,\ 30,\ 20,\ \frac{40}{3},\ \frac{80}{9},\ \frac{160}{27},\ \frac{320}{81},\ \ldots.$
19. $1,\ 3,\ 15,\ 105,\ 945,\ 10{,}395,\ 135{,}135,\ \ldots.$
20. $4,\ 8,\ 32,\ 192,\ 1{,}536,\ 15{,}360,\ \ldots.$
21. $3,\ 21,\ 231,\ 3{,}465,\ 65{,}835,\ \ldots.$
22. $\frac{5}{16},\ \frac{5}{4},\ 10,\ 120,\ 1{,}920,\ 38{,}400,\ \ldots.$
23. $5,\ 20,\ 100,\ 600,\ 4{,}200,\ 33{,}600,\ 302{,}400,\ \ldots.$

24. $2, 6, 30, 210, 1{,}890, 20{,}790, 270{,}270, \ldots.$

25. $1, 2, 8, 48, 384, 3{,}840, 46{,}080, \ldots.$

26. $\frac{1}{9}, \frac{5}{9}, \frac{35}{9}, 35, 385, 5{,}005, 75{,}075, \ldots.$

In problems 27–32, write a **formula** of each piecewise sequence:

27. $-4, 10, -16, 22, -28, 34, \ldots.$

28. $2, 4, -6, 8, 10, 12, -14, 16, \ldots.$

29. $-1, \sqrt{2}, 2, \sqrt{8}, -4, \sqrt{32}, 8, \sqrt{128}, \ldots.$

30. $-9, -12, 15, 18, -21, -24, 27, 30, \ldots.$

31. $5, 10, -15, -20, 25, 30, -35, -40, \ldots.$

32. $6, -12, -18, 24, 30, -36, -42, 48, \ldots.$

In problems 33–36, simplify each product-type sequence:

33. $\prod_{i=1}^{n} 4^i.$

34. $\prod_{i=1}^{n} 2^{2i-1}.$

35. $\prod_{i=1}^{n} \left[\frac{i}{i+2}\right].$

36. $\prod_{i=1}^{n} \left[\frac{i}{i+3}\right].$

Chapter 4

Summations and Proof by Induction

This chapter's aims are to examine summations parallel to Eqs. (3.3), (3.4), and (3.6). The additional objectives are to get acquainted with the **proof by induction** method to verify Eqs. (3.3), (3.4), and (3.6). First let's commence with the examination of linear and geometric summations.

4.1 Linear and Geometric Summations

The succeeding three examples will apply Eqs. (3.3) and (3.4) to simplify various summations. The first example will implement Eq. (3.3) to add consecutive multiples of 6.

Example 4.1. Using Eq. (3.3), simplify the following summation:

$$6 + 12 + 18 + 24 + 30 + \cdots + 300. \tag{4.1}$$

Solution: First reformulate (4.1) as:

$$6 \cdot 1 + 6 \cdot 2 + 6 \cdot 3 + 6 \cdot 4 + 6 \cdot 5 + \cdots + 6 \cdot 50. \tag{4.2}$$

Note that (4.2) is expressed in the expanded form. Next, reformulate (4.2) in the factored form or in the **sigma notation** as:

$$\sum_{i=1}^{50} 6i. \tag{4.3}$$

Thus, via Eq. (3.3) simplify (4.3) as:

$$6 \cdot \left[\frac{50 \cdot 51}{2}\right] = 150 \cdot 51.$$

The next example applies Eq. (3.3) to add consecutive terms that are three more than multiples of 4.

Example 4.2. Using Eq. (3.3), simplify the following summation:

$$7 + 11 + 15 + 19 + 23 + \cdots + 227. \tag{4.4}$$

Solution: First, reformulate (4.4) as:

$$(4 \cdot 1 + 3) + (4 \cdot 2 + 3) + (4 \cdot 3 + 3) + (4 \cdot 4 + 3) + (4 \cdot 5 + 3) + \cdots + (4 \cdot 56 + 3). \tag{4.5}$$

Next, rewrite (4.5) in the factored form or in the **sigma notation** as:

$$\sum_{i=1}^{56} (4i + 3). \tag{4.6}$$

Using the **distributive property of summations**, regroup (4.6) as:

$$\sum_{i=1}^{56} 4i + \sum_{i=1}^{56} 3 \tag{4.7}$$

Thus, via Eq. (3.3) simplify (4.7) as:

$$4 \cdot \left[\frac{56 \cdot 57}{2}\right] + 3 \cdot 56 = 112 \cdot 57 + 3 \cdot 56.$$

The succeeding example will apply Eqs. (3.3) and (3.4) to simplify an alternating summation.

Example 4.3. Using Eqs. (3.3) and (3.4), simplify the corresponding alternating summation:

$$\sum_{i=1}^{20} (-1)^i \, i. \tag{4.8}$$

Solution: First, decompose (4.8) into two summations of even integers and odd integers as:

$$\sum_{i=1}^{20} (-1)^i \, i = -1 + 2 - 3 + 4 - 5 + \cdots - 19 + 20. \tag{4.9}$$

First, notice that (4.9) adds 10 odd integers (in red) and 10 even integers (in blue). Second of all, observe that the odd integers in red have a negative sign while the even integers in blue have a positive sign. By regrouping the even integers and the odd integers in (4.9) into two separate summations and together with Eqs. (3.3) and (3.4) you obtain:

$$\begin{aligned}
&[2 + 4 + 6 + \cdots + 20] - [1 + 3 + 5 + \cdots + 19] \\
&= \sum_{i=1}^{10} 2i - \sum_{i=1}^{10} (2i - 1) \\
&= 2 \cdot \left[\frac{10 \cdot 11}{2}\right] - 10^2 \\
&= 10.
\end{aligned}$$

The upcoming two examples will apply Eq. (3.6) to simplify specific **geometric summations**. The first example will implement (3.6) to determine the geometric summation of powers of 3.

Example 4.4. Using Eq. (3.6), simplify the following geometric summation:

$$1 + 3 + 9 + 27 + 81 + 243 + 729 + 2{,}187. \tag{4.10}$$

Solution: First, reformulate (4.10) as:

$$3^0 + 3^1 + 3^2 + 3^3 + 3^4 + 3^5 + 3^6 + 3^7. \tag{4.11}$$

Note that (4.11) adds eight terms, where the starting term is $a = 1$ and the multiplicative factor is $r = 3$. Thus, via Eq. (3.6) you simplify

(4.11) as:

$$\frac{1-3^8}{1-3} = \frac{3^8-1}{3-1} = 3{,}280.$$

The next example will implement Eq. (3.6) to determine the geometric summation with powers of $\sqrt{2}$.

Example 4.5. Using Eq. (3.6), simplify the following geometric summation:

$$2+\sqrt{8}+4+\sqrt{32}+8+\sqrt{128}+16+\sqrt{512}. \tag{4.12}$$

Solution: First, reformulate (4.12) as:

$$\left[\sqrt{2}\right]^2+\left[\sqrt{2}\right]^3+\left[\sqrt{2}\right]^4+\left[\sqrt{2}\right]^5+\left[\sqrt{2}\right]^6+\left[\sqrt{2}\right]^7 \tag{4.13}$$
$$+\left[\sqrt{2}\right]^8+\left[\sqrt{2}\right]^9.$$

In (4.11), you are adding eight terms, where the starting term is $a = 2$ and the multiplicative factor is $r = \sqrt{2}$. Thus, via Eq. (3.6) simplify (4.13) as:

$$\frac{2\left[1-\left[\sqrt{2}\right]^8\right]}{1-\sqrt{2}} = \frac{2\left[\left[\sqrt{2}\right]^8-1\right]}{\sqrt{2}-1},$$

$$= \frac{30}{\sqrt{2}-1} = \frac{30}{\sqrt{2}-1}\cdot\left[\frac{\sqrt{2}+1}{\sqrt{2}+1}\right] = 30\cdot[\sqrt{2}+1].$$

4.2 Proof by Induction

This section's aims are to prove Eqs. (3.3), (3.4), and (3.6) by applying the **proof by induction** method. In addition, the objectives are to derive supplemental summations using Eqs. (3.3), (3.4) and (3.6). The first example will apply the proof by induction method to prove (3.3).

Example 4.6. Using **proof by induction**, verify that the following summation holds true:

$$1+2+3+4+\cdots+(n-1)+n=\sum_{i=1}^{n} i=\frac{n\cdot[n+1]}{2}. \tag{4.14}$$

Solution: First, note that (4.14) holds true for $n=3$ as:

$$1+2+3 = \frac{3\cdot 4}{2}=6.$$

Next, assume that (4.14) holds true for $n=k$:

$$1+2+3+4+\cdots+(k-1)+k=\sum_{i=1}^{k} i=\frac{k\cdot[k+1]}{2}. \tag{4.15}$$

Now verify that (4.15) holds true for $n=k+1$:

$$[1+2+3+4+\cdots+(k-1)+k]+[k+1]=\sum_{i=1}^{k+1} i=\frac{[k+1]\cdot[k+2]}{2}. \tag{4.16}$$

Using (4.15), reformulate (4.16) as:

$$\left[\sum_{i=1}^{k} i\right]+[k+1]=\left[\frac{k\cdot[k+1]}{2}\right]+[k+1]=\frac{[k+1]\cdot[k+2]}{2}.$$

Hence, the result follows.

The upcoming example will apply the **proof by induction** method to prove (3.4).

Example 4.7. Using **proof by induction**, verify that the following summation holds true:

$$1+3+5+7+\cdots+(2n-1)=\sum_{i=1}^{n}(2i-1)=n^2. \tag{4.17}$$

Solution: Observe that (4.17) holds true for $n=4$ as:

$$1+3+5+7=16=4^2.$$

Now, assume that (4.17) holds true for $n = k$:

$$1 + 3 + 5 + 7 + \cdots + (2k - 1) = \sum_{i=1}^{k} (2i - 1) = k^2. \tag{4.18}$$

Next, verify that (4.18) holds true for $n = k + 1$:

$$[1 + 3 + 5 + 7 + \cdots + (2k - 1)] + [2k + 1] = \sum_{i=1}^{k+1} (2i - 1) = (k + 1)^2. \tag{4.19}$$

Using (4.18), reformulate (4.19) as:

$$\left[\sum_{i=1}^{k} i\right] + [2k + 1] = \left[k^2\right] + [2k + 1] = (k + 1)^2.$$

The result follows.

The consequent example will apply the **proof by induction** method to prove a special case of (3.6).

Example 4.8. Using **proof by induction**, verify that the following summation holds true:

$$1 + 2^1 + 2^2 + 2^3 + \cdots + 2^{n-1} + 2^n = \sum_{i=0}^{n} 2^i = 2^{n+1} - 1. \tag{4.20}$$

Solution: Observe that (4.20) holds true for $n = 2$ as:

$$1 + 2^1 + 2^2 = 7 = 2^3 - 1.$$

Now assume that (4.20) holds true for $n = k$:

$$1 + 2^1 + 2^2 + 2^3 + \cdots + 2^{k-1} + 2^k = \sum_{i=0}^{k} 2^i = 2^{k+1} - 1. \tag{4.21}$$

Now confirm that (4.21) holds true for $n = k + 1$:

$$\left[1 + 2^1 + 2^2 + 2^3 + \cdots + 2^{k-1} + 2^k\right] + \left[2^{k+1}\right] = \sum_{i=0}^{k+1} 2^i = 2^{k+2} - 1. \tag{4.22}$$

Using (4.21), reformulate (4.22) as:

$$\left[\sum_{i=0}^{k} 2^i\right] + \left[2^{k+1}\right] = \left[2^{k+1} - 1\right] + 2^{k+1} = 2 \cdot 2^{k+1} - 1 = 2^{k+2} - 1.$$

Hence, the result follows.

Example 4.9. Using **proof by induction**, verify that the following summation holds true:

$$5 + 9 + 13 + 17 + \cdots + (4n-3) + (4n+1) = \sum_{i=1}^{n} (4i+1) = n \cdot [2n+3]. \tag{4.23}$$

Solution: Observe that (4.23) holds true for $n = 4$ as:

$$5 + 9 + 13 + 17 = 44 = 4[2 \cdot 4 + 3] = 4 \cdot 11.$$

Now assume that (4.23) holds true for $n = k$:

$$5+9+13+17+\cdots+(4k-3)+(4k+1) = \sum_{i=1}^{k}(4i+1) = k \cdot [2k+3]. \tag{4.24}$$

Now confirm that (4.24) holds true for $n = k + 1$:

$$\begin{aligned} &[5 + 9 + 13 + 17 + \cdots + (4k-3) + (4k+1)] + [4k+5] \\ &= \sum_{i=1}^{k+1} (4i+1) = [k+1] \cdot [2k+5]. \end{aligned} \tag{4.25}$$

Using (4.24), reformulate (4.25) as:

$$\begin{aligned} &\left[\sum_{i=1}^{k} (4i+1)\right] + [4k+5] \\ &= k \cdot [2k+3] + [4k+5] = 2k^2 + 7k + 5 = [k+1] \cdot [2k+5]. \end{aligned}$$

Hence, the result follows.

The next sequence of examples will apply Eq. (3.3) to derive specific summations. The first example adds consecutive positive even integers.

Example 4.10. Using (3.3), simplify the following finite summation:

$$2 + 4 + 6 + 8 + 10 + 12 + \cdots . \tag{4.26}$$

Solution: For $n \in \mathbb{N}$, reformulate (4.26) in the sigma notation as:

$$2 + 4 + 6 + 8 + 10 + 12 + \cdots + 2n = \sum_{i=1}^{n} 2i. \tag{4.27}$$

Note that (4.27) adds multiples of 2. Next, factor the 2 and reformulate (4.27) as:

$$2 \cdot [1 + 2 + 3 + 4 + \cdots + n] = 2 \cdot \left[\sum_{i=1}^{n} i\right] = n \cdot [n + 1]. \tag{4.28}$$

Proving (4.28) by induction will be left as an end of chapter exercise.

The upcoming example adds positive integers that are one less than multiples of 3.

Example 4.11. Using Eq. (3.3), simplify the following finite summation:

$$2 + 5 + 8 + 11 + 14 + 17 + \cdots . \tag{4.29}$$

Solution: For $n \in \mathbb{N}$, reformulate (4.29) in the sigma notation as:

$$2 + 5 + 8 + 11 + 14 + 17 \ + \cdots + (3n - 1) = \sum_{i=1}^{n} (3i - 1). \tag{4.30}$$

Note that (4.30) adds integers that are one less than a multiple of 3. Next, by using the **distributive property** of summations, you decompose (4.30) and acquire:

$$\sum_{i=1}^{n} (3i-1) = \sum_{i=1}^{n} 3i - \sum_{i=1}^{n} 1 = \frac{3n \cdot [n+1]}{2} - n = \frac{n \cdot [3n+1]}{2}. \tag{4.31}$$

Proving (4.31) by induction will be left as an end of chapter exercise.

The succeeding example adds consecutive positive integers starting with 7.

Example 4.12. Using Eq. (3.3), derive the formula of the following summation:

$$7+8+9+10+11+12+13+\cdots+n. \tag{4.32}$$

Solution: For $n \geq 7$, rewrite (4.32) in the sigma notation as:

$$7+8+9+10+11+12+13+\cdots+n=\sum_{i=7}^{n} i. \tag{4.33}$$

Next, by downshifting the starting and the terminating indices of (4.33) by 6, for $n \geq 7$ you obtain the corresponding summation:

$$\sum_{i=7}^{n} i=\sum_{i=1}^{n-6}(i+6). \tag{4.34}$$

By implementing the **distributive property** of summations, decompose (4.34) into two separate summations and for all $n \geq 7$ you procure:

$$\sum_{i=1}^{n-6}(i+6)=\sum_{i=1}^{n-6} i+\sum_{i=1}^{n-6} 6=\frac{[n-6]\cdot[n-5]}{2}+6\,[n-6]$$
$$=\frac{[n-6]\cdot[n+7]}{2}.$$

From Example (4.12), for all $m \in [2,\ldots,n]$ you can then extend (4.33) to the related summation of specific consecutive positive integers starting with m:

$$m+(m+1)+(m+2)+(m+3)+(m+4)+\cdots+n=\sum_{i=m}^{n} i. \tag{4.35}$$

Obtaining a summation formula of (4.35) will be left as an end of chapter exercise.

4.3 Exercises

In problems 1 − 4, use (3.3) and the **distributive property** to determine the following summations:

1. $3+6+9+12+15+\cdots+240$.
2. $4+8+12+16+20+\cdots+320$.
3. $5+9+13+17+21+\cdots+81$.

4. $2 + 5 + 8 + 11 + 14 + \cdots + 92.$

In problems 5 – 8, use (3.6) to determine the following summations:

5. $\frac{1}{4} + \frac{1}{2} + 1 + 2 + 4 + \cdots + 128.$
6. $4 + 2\sqrt{2} + 2 + \sqrt{2} + 1 + \frac{1}{\sqrt{2}} + \cdots + \frac{1}{8}.$
7. $\frac{1}{9} - \frac{1}{3} + 1 - 3 + 9 - \cdots + 729.$
8. $1 - \frac{\sqrt{3}}{2} + \frac{3}{4} - \frac{3\sqrt{3}}{8} + \frac{9}{16} - \cdots + \frac{243}{1024}.$

In problems 9 – 18, prove the following expressions by the **proof by induction** method:

9. $\sum_{i=1}^{k} (2i - 1) = k^2.$
10. $\sum_{i=1}^{k} (4i - 3) = k \cdot [2k - 1].$
11. $\sum_{i=1}^{k} i^2 = \frac{k \cdot [k+1] \cdot [2k+1]}{6}.$
12. $\sum_{i=1}^{k} i[i + 1] = \frac{k \cdot [k+1] \cdot [k+2]}{3}.$
13. $\sum_{i=0}^{k} a \cdot r^i = \frac{a \cdot [1 - r^{k+1}]}{1 - r}, \quad (r \neq 1).$
14. $\sum_{i=1}^{k} i \cdot [i + 1] \cdot [i + 2] = \frac{k \cdot [k+1] \cdot [k+2] \cdot [k+3]}{4}.$
15. $\sum_{i=1}^{k} \frac{1}{i \cdot [i+1]} = \frac{k}{k+1}.$
16. $\sum_{i=1}^{k} \frac{1}{[2i-1] \cdot [2i+1]} = \frac{k}{2k+1}.$
17. $\sum_{i=1}^{k} i \cdot 2^{i-1} = [k - 1] \cdot 2^k + 1.$
18. $\sum_{i=1}^{k} i \cdot i! = (k + 1)! \; - \; 1.$

In problems 19 – 22, use (3.3) and the **distributive property** to derive the following summations:

19. $\sum_{i=7}^{k} i, \quad (i \geq 7).$
20. $\sum_{i=8}^{k} i, \quad (i \geq 8).$
21. $\sum_{i=9}^{k} i, \quad (i \geq 9).$
22. Using Exercises 19 – 21, determine the formula of (4.35).

Chapter 5

Traits of Pascal's Triangle

Chapter 1 introduced the **Pascal's triangle** presented in Figure 5.1.

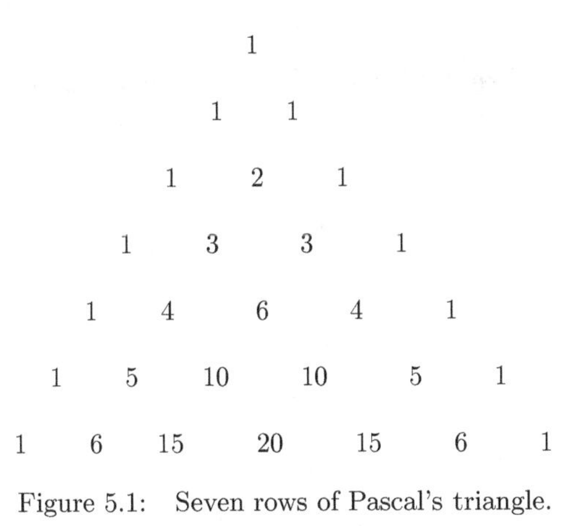

Figure 5.1: Seven rows of Pascal's triangle.

This chapter's aims are to derive various traits of the Pascal's triangle and prove them by applying the definition of combinations (5.1) and together with the proof by induction method. You will examine Figure 5.1 in the horizontal direction to analyze the triangle's **horizontal identities** and in the diagonal direction to examine the triangle's **diagonal identities**. To obtain the triangle's horizontal identities, you will restructure Figure 5.1 into blue and red horizontal rows. The blue rows are even-ordered rows and the red rows are odd-ordered rows presented in Figure 5.2.

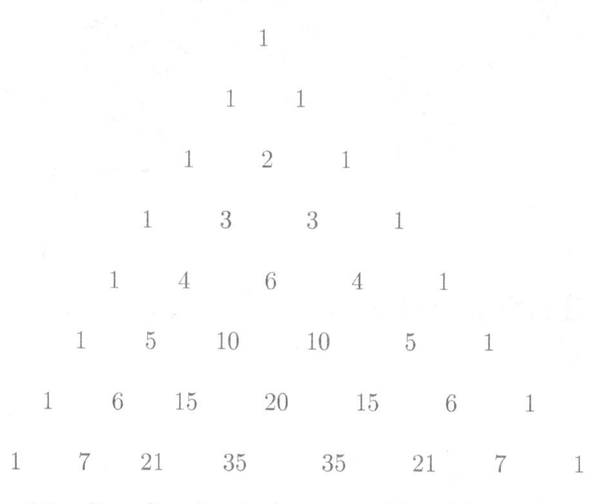

Figure 5.2: Pascal's triangle decomposed into blue and red rows.

Analogously, to acquire the triangle's diagonal identities, you will restructure Figure 5.1 into blue and red diagonals as indicated in Figure 5.3.

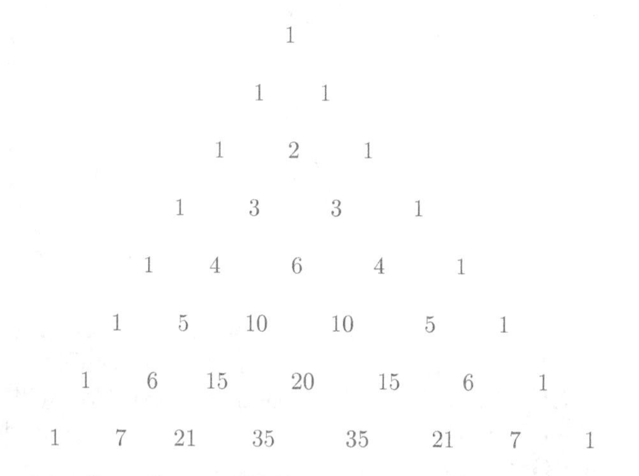

Figure 5.3: Pascal's triangle decomposed into blue and red diagonals.

Next using the definition of combinations (5.1), you reformulate Figure 5.1 resembled in Figure 5.4.

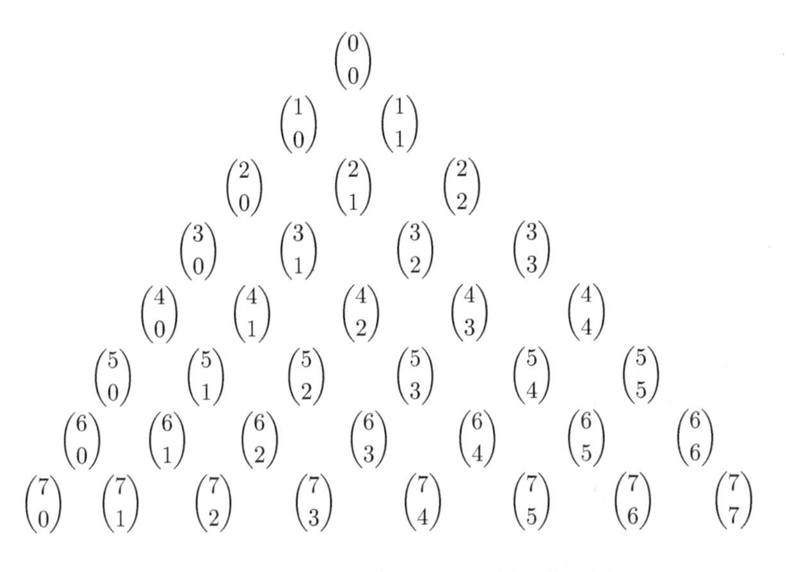

Figure 5.4: Pascal's triangle expressed in Combinations.

Figure 5.4 guides you to the definition of **combinations** expressed in terms of factorials.

Definition 10. For all $n \geq 0$ and $k \in [0, 1, \ldots, n]$, the number of k-combinations out of n elements is defined as the corresponding binomial coefficient:

$$\binom{n}{k} = \frac{n!}{k!(n-k)!}. \tag{5.1}$$

Note that for $k \in [0, 1, \ldots, n]$, (5.1) characterizes the number of k-combinations out of the n elements. Throughout this chapter, you will apply (5.1) to determine and describe several features of the Pascal's triangle by coalescing Figure 5.2 together with Figure 5.4. First, you will examine the triangle's horizontal rows that lead to the **horizontal identities**. In addition, you will combine Figure 5.3 together with Figure 5.4 that will decipher the triangle's diagonals that will direct to the **diagonal identities**.

Next, the objective is to present the applications of (5.1) to determine the triangle's elements in the third row. Figure 5.2 restructures the Pascal's triangle into blue and red rows as shown in the corresponding sketch as follows:

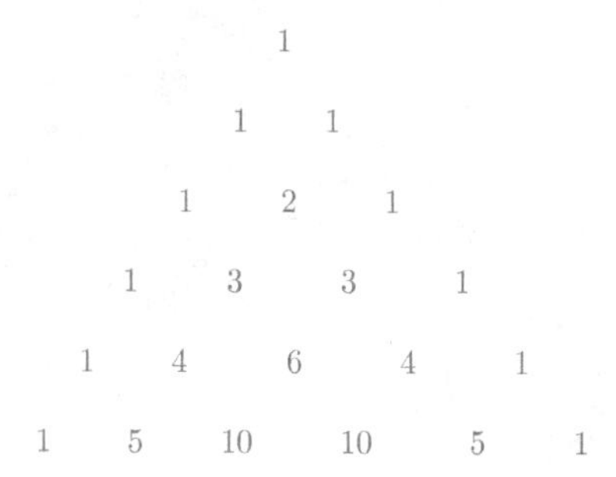

From Figure 5.4, the aim is to compute each element of Pascal's triangle by applying (5.1). For instance, by applying (5.1) you obtain all the elements of the triangle's third row as follows:

$$\binom{3}{0} = \frac{3!}{0!3!} = 1, \quad \binom{3}{1} = \frac{3!}{1!2!} = 3, \quad \binom{3}{2} = \frac{3!}{2!1!} = 3,$$

$$\binom{3}{3} = \frac{3!}{3!0!} = 1.$$

The next aim is to apply (5.1) to enumerate all the combinations 2 out of 5 from the set $\{a, b, c, d, e\}$ and express all the combinations in terms of (5.1) and (3.3).

Next restructure all the possible combinations in four rows in the following **upper-triangular configuration** represented in (5.2):

$$\begin{array}{rrrr} \{a, b\}, & \{a, c\}, & \{a, d\}, & \{a, e\} \\ & \{b, c\}, & \{b, d\}, & \{b, e\} \\ & & \{c, d\}, & \{c, e\} \\ & & & \{d, e\}. \end{array} \tag{5.2}$$

By adding all the terms from each row in (5.2) you procure the corresponding summation:

$$1 + 2 + 3 + 4 = 10 = \frac{4 \cdot 5}{2} = \binom{5}{2} = \frac{5!}{2!3!}. \tag{5.3}$$

Therefore, for all $n \in \mathbb{N}$ you can extend (5.3) to the corresponding result:

$$\sum_{i=1}^{n} i = \frac{n \cdot [n+1]}{2} = \binom{n+1}{2}. \tag{5.4}$$

Equation (5.4) will emerge as one of the triangle's properties in Section 5.2 and direct you to supplemental diagonal identities. Section 5.1 will examine the triangle's horizontal identities.

5.1 Horizontal Identities

This section will focus on the triangle's **horizontal identities** by applying Figure 5.2 together with Figure 5.4. The section will first commence with the triangle's **Symmetry identity**.

Figure 5.5 describes the triangle's **Symmetry pattern** with the blue, green and red colors.

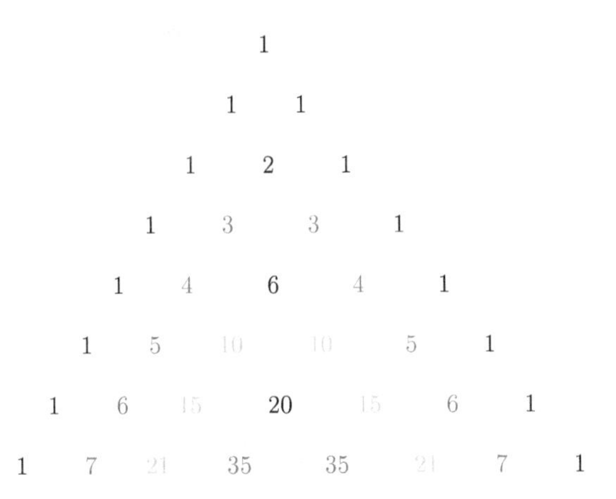

Figure 5.5: Pascal's triangle symmetry property.

Via Figure (5.5) and via (5.1), the Triangle's blue terms resemble the following configuration:

$$\begin{array}{ccccc} 3 = 3, & 4 = 4, & 5 = 5, & 6 = 6, & 7 = 7 \\ \binom{3}{1} = \binom{3}{2}, & \binom{4}{1} = \binom{4}{3}, & \binom{5}{1} = \binom{5}{4}, & \binom{6}{1} = \binom{6}{5}, & \binom{7}{1} = \binom{7}{6}. \end{array} \tag{5.5}$$

Note from (5.5) you acquire: $1 + 2 = 3$, $1 + 3 = 4$, $1 + 4 = 5$, and $1 + 5 = 6$. In addition, via (5.5) and (5.1) you obtain:

$$\binom{3}{1} = \frac{3!}{1!2!} = \binom{3}{2}, \quad \binom{4}{1} = \frac{4!}{1!3!} = \binom{4}{3}, \quad \binom{5}{1} = \frac{5!}{4!1!} = \binom{5}{4}.$$

Next, observe that the triangle's green terms resemble the corresponding structure:

$$\begin{aligned} 10 &= 10, & 15 &= 15, & 21 &= 21, \\ \binom{5}{2} &= \binom{5}{3}, & \binom{6}{2} &= \binom{6}{4}, & \binom{7}{2} &= \binom{7}{5}. \end{aligned} \tag{5.6}$$

Via (5.6) you procure: $2+3=5$, $1+3=4$, $2+4=6$, and $2+5=7$. Via (5.6) and (5.1) you get:

$$\binom{5}{2} = \frac{5!}{2!3!} = \binom{5}{3}, \quad \binom{6}{2} = \frac{6!}{2!4!} = \binom{6}{4}, \quad \binom{7}{2} = \frac{7!}{2!5!} = \binom{7}{5}, \quad \ldots.$$

Next, note that the triangle's red terms depict the corresponding pattern:

$$\begin{aligned} 35 &= 35, \\ \binom{7}{3} &= \binom{7}{4}, \end{aligned} \tag{5.7}$$

where $3+4=7$. Via (5.7) and (5.1) you obtain:

$$\binom{7}{3} = \frac{7!}{3!4!} = \binom{7}{4}.$$

Hence via (5.5), (5.6) and (5.7), for all $n \geq 0$ and $k = 0, 1, \ldots, n$ you procure:

$$\binom{n}{k} = \binom{n}{n-k}, \tag{5.8}$$

where $k + (n-k) = n$ for all $k = 0, 1, \ldots, n$. Thus via (5.1) you obtain:

$$\binom{n}{n-k} = \frac{n!}{(n-k)![n-(n-k)]!} = \frac{n!}{(n-k)!k!} = \binom{n}{k}.$$

Hence, the result follows.

The next aim is to examine the triangle's **Pascal's identity** by adding two neighboring horizontal terms in each row. Figure 5.6 presents the **Pascal's identity** with blue and red colors, where the each red term is the sum of the two adjacent horizontal blue terms.

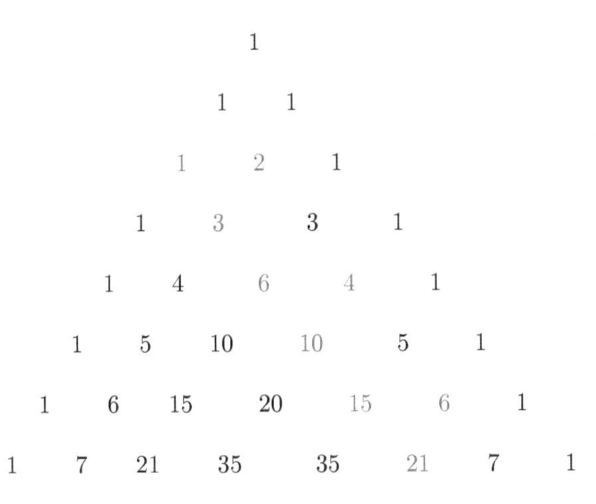

Figure 5.6: Representation of the Pascal's identity.

When adding two adjoining horizontal blue terms in each row in Figure 5.6 together with (5.1), you procure the following pattern:

$$1 + 2 = 3, \quad 6 + 4 = 10, \quad 15 + 6 = 21, \ldots$$

$$\binom{2}{0} + \binom{2}{1} = \binom{3}{1}, \binom{4}{2} + \binom{4}{3} = \binom{5}{3}, \binom{6}{4} + \binom{6}{5} = \binom{7}{5}, \ldots \tag{5.9}$$

By applying (5.9) together with (5.1), you acquire:

$$\binom{6}{4} + \binom{6}{5} = \frac{6!}{2!4!} + \frac{6!}{5!1!} = \frac{6!5}{2!5!} + \frac{6!2}{5!2!} = \frac{6![5+2]}{5!2!} = \frac{7!}{5!2!} = \binom{7}{5}.$$

In addition via (5.9), for all $n \in \mathbb{N}$ and $k \in [0, 1, \ldots, n-1]$ you obtain the corresponding **Pascal's identity**:

$$\binom{n}{k} + \binom{n}{k+1} = \binom{n+1}{k+1}. \tag{5.10}$$

Proving Eq. (5.10) (**Pascal's identity**) will be left as an end of the chapter exercise. Equation (5.10) will also be used to prove supplemental identities.

In Figure 5.6 you added two horizontal neighboring terms in each row and obtained the **Pascal's identity**. Next, you will affix all the horizontal terms in each row and produce the Triangle's **Power**

identity (adding up to powers of 2). Figure 5.7 decomposes the triangle's rows, where the blue terms resemble the even-ordered rows while the red terms portray the odd-ordered rows.

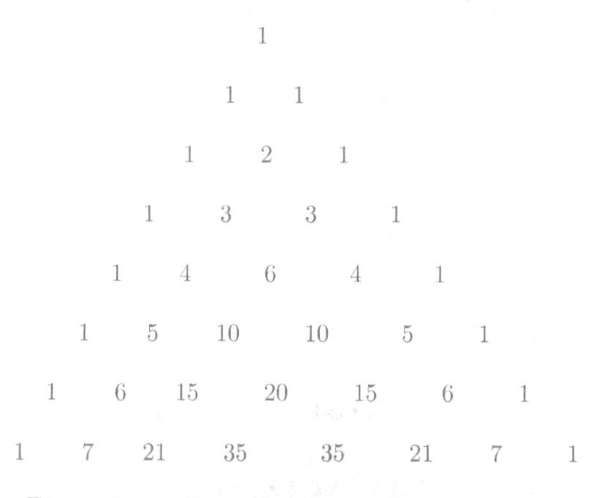

Figure 5.7: Pascal's triangle Power identity.

Now you will combine all the terms in each row. Via Figure 5.7, by combining all the terms in each row starting with the 0th row and by applying (5.1) you obtain the following properties:

$$\begin{aligned}
1 &= \binom{0}{0} = 2^0 \quad \text{(0th row)}, \\
1+1 &= \binom{1}{0} + \binom{1}{1} = 2^1 \quad \text{(1st row)}, \\
1+2+1 &= \binom{2}{0} + \binom{2}{1} + \binom{2}{2} = 2^2 \quad \text{(2nd row)}, \\
1+3+3+1 &= \binom{3}{0} + \binom{3}{1} + \binom{3}{2} + \binom{3}{3} = 2^3 \quad \text{(3rd row)}, \\
&\vdots
\end{aligned} \tag{5.11}$$

Note that the power of 2 corresponds directly to the order of each row. Hence for $n \in \mathbb{N}$, (5.11) extends to the associated **Power identity**:

$$\sum_{i=0}^{n} \binom{n}{i} = 2^n. \tag{5.12}$$

Note that the blue even-ordered rows have an odd number of terms, while the red odd-ordered rows have an even number of terms. Thus, proving (5.12) will require two cases when n is even and when n is odd. In addition, (5.12) is proved by induction by applying the **Symmetry identity** and the **Pascal's identity**. Furthermore, from Figure 5.7, and via (5.11) and (5.12), by adding all the elements of the Pascal's triangle you procure:

$$2^0 + 2^1 + 2^1 + \cdots + 2^n = \sum_{i=0}^{n} 2^i = 2^{n+1} - 1.$$

Equation (5.12) will guide you to further special cases and applications of the Power identity.

5.1.1 *Additional Horizontal Identities*

This section's objectives are to explore deeper attributes of the Pascal's triangle together with the Power identity (5.12). The Power identity is derived by adding all the terms in each row.

Figure 5.8 will focus on the special case of (5.12) by adding the first half of the terms in odd-ordered rows starting with the third row.

Figure 5.8: Pascal's triangle Power identity in odd-ordered rows.

First you combine the first half of the terms in rows 3, 5 and 7 together with (5.1) and hence formulate the following characteristics:

$$1+3=\binom{3}{0}+\binom{3}{1}=4=4^1 \text{ (3rd row)},$$

$$1+5+10=\binom{5}{0}+\binom{5}{1}+\binom{5}{2}=16=4^2 \text{ (5th row)},$$

$$1+7+21+35=\binom{7}{0}+\binom{7}{1}+\binom{7}{2}+\binom{7}{3}=64=4^3 \text{ (7th row)},$$

$$\vdots \tag{5.13}$$

Thus, for $n \in \mathbb{N}$, (5.13) transitions to the corresponding modification of the Power identity:

$$\sum_{i=0}^{n} \binom{2n+1}{i} = 4^n. \tag{5.14}$$

Power identity (5.12) will be applied to prove (5.14) and will be left as an end of chapter exercise.

The next aim is to apply the Power identity (5.12) together with the Symmetry identity (5.8) as well as (5.13). Analogous to Example (3.13), you will implement the lower-triangular matrix grouping of the terms.

Using (5.12), (5.8) and (5.13), for $n = 3$ and $n = 4$ verify the corresponding formula:

$$\sum_{i=0}^{n} [i+1]\binom{n}{i} = [n+2]2^{n-1}. \tag{5.15}$$

Solution: First, for $n = 3$, you will analyze the triangle's third row and verify the cognate version of (5.15):

$$\sum_{i=0}^{3} [i+1]\binom{3}{i} = 5\cdot 2^2. \tag{5.16}$$

Parallel to Example (3.13), arrange the terms of (5.16) in corresponding lower-triangular matrix form:

$$\begin{array}{l} \binom{3}{0} \\ \binom{3}{1} + \binom{3}{1} \\ \binom{3}{2} + \binom{3}{2} + \binom{3}{2} \\ \binom{3}{3} + \binom{3}{3} + \binom{3}{3} + \binom{3}{3}. \end{array} \tag{5.17}$$

Next, reformulate the binomial terms of (5.17) by colors as follows:

$$\left[\binom{3}{0} + \binom{3}{1} + \binom{3}{2} + \binom{3}{3}\right] + \left[\binom{3}{3} + \binom{3}{1} + \binom{3}{2} + \binom{3}{3}\right] + \left[\binom{3}{2} + \binom{3}{3}\right]. \tag{5.18}$$

By using the Symmetry identity (5.8) you obtain:

$$\binom{3}{3} = \binom{3}{0}, \quad \binom{3}{2} = \binom{3}{1} \quad \text{and} \quad \binom{3}{3} = \binom{3}{0},$$

and then reformulate (5.18) as:

$$\left[\binom{3}{0} + \binom{3}{1} + \binom{3}{2} + \binom{3}{3}\right] + \left[\binom{3}{0} + \binom{3}{1} + \binom{3}{2} + \binom{3}{3}\right] + \left[\binom{3}{1} + \binom{3}{0}\right]. \tag{5.19}$$

Using (5.12) and (5.13), (5.19) simplifies to the corresponding powers of 2:

$$2^3 + 2^3 + 4^1 = 2 \cdot 2^2 + 2 \cdot 2^2 + 2^2 = 5 \cdot 2^2.$$

The result follows. Similarly, for $n = 4$ you will analyze the triangle's fourth row and verify the following version of (5.15):

$$\sum_{i=0}^{4} [i + 1] \binom{4}{i} = 6 \cdot 2^3. \tag{5.20}$$

Mimicking (5.17), restructure the terms of (5.20) in cognate lower-triangular matrix form:

$$\begin{aligned}
&\binom{4}{0}\\
&\binom{4}{1} + \binom{4}{1}\\
&\binom{4}{2} + \binom{4}{2} + \binom{4}{2}\\
&\binom{4}{3} + \binom{4}{3} + \binom{4}{3} + \binom{4}{3}\\
&\binom{4}{4} + \binom{4}{4} + \binom{4}{4} + \binom{4}{4} + \binom{4}{4}.
\end{aligned} \tag{5.21}$$

Next, reformulate the binomial terms of (5.21) by colors as follows:

$$\begin{aligned}
&\left[\binom{4}{0} + \binom{4}{1} + \binom{4}{2} + \binom{4}{3} + \binom{4}{4}\right]\\
&\quad + \left[\binom{4}{4} + \binom{4}{1} + \binom{4}{2} + \binom{4}{3} + \binom{4}{4}\right]\\
&\quad + \left[\binom{4}{4} + \binom{4}{3} + \binom{4}{2} + \binom{4}{3} + \binom{4}{4}\right].
\end{aligned} \tag{5.22}$$

Using the Symmetry identity (5.8) and the Power identity (5.12), (5.22) simplifies to the corresponding powers of 2:

$$2^4 + 2^4 + 2^4 = 2 \cdot 2^3 + 2 \cdot 2^3 + 2 \cdot 2^3 = 6 \cdot 2^3.$$

The result follows.

Section 5.2 will focus on the Triangle's **diagonal identities.**

5.2 Diagonal Identities

This section will focus on the triangle's **diagonal identities** by applying Figure 5.3 together with Figure 5.4. The succeeding examples will combine the adjacent diagonal terms in comparison to combining horizontal terms as we did in the previous section. In Figure 5.6, you added two horizontal neighbors and obtained the

Pascals' identity. Analogously, you will initiate the **Square identity** by adding two neighboring terms in the second diagonal.

The **Square identity** combines two neighboring blue diagonal terms in the second diagonal as illustrated in Figure 5.9:

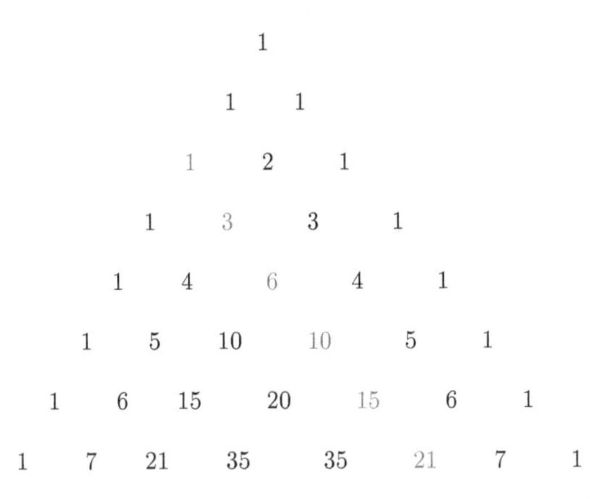

Figure 5.9: The Square identity and the triangle's second diagonal.

You procure the following identities by expressing the sum of two adjacent blue terms in the second diagonal in Figure 5.9 together with (5.1):

$$\begin{aligned} 1+3 &= \binom{2}{0} + \binom{3}{1} = 2^2, \\ 3+6 &= \binom{3}{1} + \binom{4}{2} = 3^2, \\ 6+10 &= \binom{4}{2} + \binom{5}{3} = 4^2, \\ 10+15 &= \binom{5}{3} + \binom{6}{4} = 5^2, \\ &\vdots \end{aligned} \tag{5.23}$$

Hence from (5.23), for all $n \geq 2$ you acquire the following **Square identity**:

$$\binom{n}{n-2} + \binom{n+1}{n-1} = n^2. \tag{5.24}$$

Verifying (5.24) will be left as an end of chapter exercise.

You will next resemble (3.3) as one of the properties of the Pascal's triangle by summing all the consecutive diagonal terms in the first diagonal. Figure 5.10 presents the red terms of the triangle's first diagonal and lists all the consecutive positive integers.

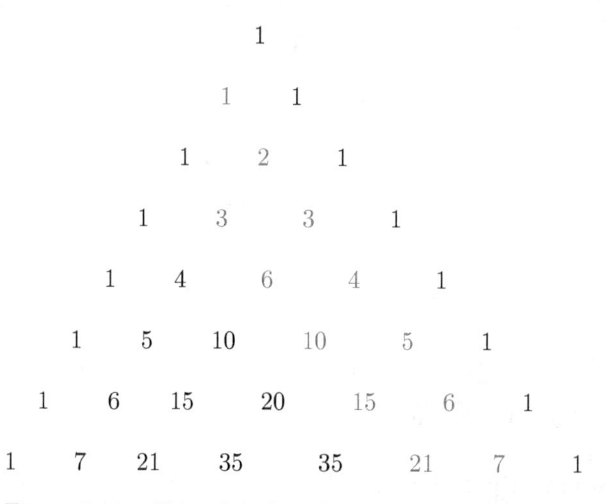

Figure 5.10: Triangle's first diagonal indicated in red.

You procure the following relations by summing all the consecutive red terms in the **first diagonal** in Figure 5.10 together with (5.1):

$$\begin{aligned}
&1+2=\binom{1}{0}+\binom{2}{1}=\binom{3}{1}=3,\\
&1+2+3=\binom{1}{0}+\binom{2}{1}+\binom{3}{2}=\binom{4}{2}=6,\\
&1+2+3+4=\binom{1}{0}+\binom{2}{1}+\binom{3}{2}+\binom{4}{3}=\binom{5}{3}=10,\\
&1+2+3+4+5=\binom{1}{0}+\binom{2}{1}+\binom{3}{2}+\binom{4}{3}+\binom{5}{4}=\binom{6}{4}=15,\\
&\vdots
\end{aligned} \tag{5.25}$$

From the relations in (5.25), for all $n \geq 2$ reformulate (3.3) as:

$$\binom{1}{0} + \binom{2}{1} + \binom{3}{2} + \binom{4}{3} + \cdots + \binom{n}{n-1} = \sum_{i=0}^{n-1} \binom{i+1}{i} = \binom{n+1}{n-1}. \tag{5.26}$$

You will prove (5.26) by induction together with the **Pascal's identity** and will be left as an end of chapter exercise.

Similar to (5.26) in Figure 5.10, by combining all the terms in the Triangle's **second diagonal** you obtain:

$$\binom{2}{0} + \binom{3}{1} + \binom{4}{2} + \binom{5}{3} + \cdots + \binom{n+1}{n-1} = \sum_{i=0}^{n-1} \binom{i+2}{i} = \binom{n+2}{n-1}. \tag{5.27}$$

By summing all terms in the triangle's **third diagonal** you procure:

$$\binom{3}{0} + \binom{4}{1} + \binom{5}{2} + \binom{6}{3} + \cdots + \binom{n+2}{n-1} = \sum_{i=0}^{n-1} \binom{i+3}{i} = \binom{n+3}{n-1}. \tag{5.28}$$

By adding all the terms in the triangle's **fourth diagonal** you acquire:

$$\binom{4}{0} + \binom{5}{1} + \binom{6}{2} + \binom{7}{3} + \cdots + \binom{n+3}{n-1} = \sum_{i=0}^{n-1} \binom{i+4}{i} = \binom{n+4}{n-1}. \tag{5.29}$$

Therefore, for all $k \in \mathbb{N}$, by affixing all the terms in the k**th diagonal**, via (5.26)–(5.29) you produce the corresponding identity:

$$\binom{k}{0} + \binom{k+1}{1} + \binom{k+2}{2} + \cdots + \binom{n+k-1}{n-1} = \sum_{i=0}^{n-1} \binom{i+k}{i} = \binom{n+k}{n-1}. \tag{5.30}$$

Proving (5.30) by induction will be left as an end of chapter exercise.

5.3 Binomial Expansion

In this section, for $n \geq 2$ the objective is to derive the binomial expansion for $(x+y)^n$ and show the correspondence with the binomial coefficients of the nth row of the Pascal's triangle. First, when $n = 2$, using the distributive property you obtain:

$$\begin{aligned}(x+y)^2 &= x^2 + 2xy + y^2, \\ &= \binom{2}{0}x^2 + \binom{2}{1}xy + \binom{2}{2}y^2.\end{aligned} \tag{5.31}$$

The coefficient of the powers of x and y in (5.31) match with the binomial coefficients of the 2nd row of the Pascal's triangle. When $n = 3$, using (5.31) together with the distributive property you get:

$$\begin{aligned}(x+y)^3 &= (x+y)^2 \cdot (x+y), \\ &= (x^2 + 2xy + y^2) \cdot (x+y), \\ &= x^3 + 3x^2y + 3xy^2 + y^3, \\ &= \binom{3}{0}x^3 + \binom{3}{1}x^2y + \binom{3}{2}xy^2 + \binom{3}{3}y^3.\end{aligned} \tag{5.32}$$

The coefficient of the powers of x and y in (5.32) render the binomial coefficients of the 3rd row of the Pascal's triangle. When $n = 4$, applying (5.32) together with the distributive property you procure:

$$\begin{aligned}(x+y)^4 &= (x+y)^3 \cdot (x+y), \\ &= (x^3 + 3x^2y + 3xy^2 + y^3) \cdot (x+y), \\ &= x^4 + 4x^3y + 6x^2y^2 + 4xy^3 + y^4, \\ &= \binom{4}{0}x^4 + \binom{4}{1}x^3y + \binom{4}{2}x^2y^2 + \binom{4}{3}xy^3 + \binom{4}{4}y^4.\end{aligned} \tag{5.33}$$

The coefficient of the powers of x and y in (5.33) mimic the binomial coefficients of the 4th row of the Pascal's triangle. Therefore for $n \geq 2$, (5.31), (5.32) and (5.33) generalize to the corresponding binomial expansion:

$$(x+y)^n = \sum_{i=0}^{n} \binom{n}{i} x^{n-i} y^i. \tag{5.34}$$

Note that in (5.34), the powers of x descend by 1, the powers of y ascend by 1, and the sum of the powers of x and y add up to n for all $i \in [0, 1, 2, \ldots, n]$. Equation (5.34) is an example of a **convolution.** Observe that (5.34) extends to the corresponding **alternating binomial expansion**:

$$(x-y)^n = \sum_{i=0}^{n} (-1)^i \binom{n}{i} x^{n-i} y^i. \tag{5.35}$$

The upcoming examples will resemble the use of (5.34).

Example 5.1. Determine the binomial expansion of:

$$(a^2+b^2)^6. \tag{5.36}$$

Solution: Applying (5.34) with $n = 6$ you get

$$(x+y)^6 = \sum_{i=0}^{6} \binom{6}{i} x^{6-i} y^i. \tag{5.37}$$

Next, substitute $x = a^2$ and $y = b^2$ into (5.37) and obtain:

$$\begin{aligned}(a^2+b^2)^6 &= \sum_{i=0}^{6} \binom{6}{i} (a^2)^{6-i} (b^2)^i, \\ &= \sum_{i=0}^{6} \binom{6}{i} a^{12-2i} b^{2i}.\end{aligned}$$

Note that the powers of a descend by 2 while the powers of b ascend by 2. When $i = 2$ you get $a^8 b^4$ with the corresponding **binomial coefficient** $\binom{6}{2}$.

Example 5.2. Determine the Binomial Expansion of:

$$\left(x + x^{-1}\right)^8. \tag{5.38}$$

Solution: Applying (5.34) with $n = 8$ you obtain:

$$(x + y)^8 = \sum_{i=0}^{8} \binom{8}{i} x^{8-i} y^i. \tag{5.39}$$

Next, substitute $y = x^{-1}$ into (5.39) and you procure:

$$\begin{aligned}(x + x^{-1})^8 &= \sum_{i=0}^{8} \binom{8}{i} (x)^{8-i} \left(x^{-1}\right)^i, \\ &= \sum_{i=0}^{8} \binom{8}{i} x^{8-i} x^{-i} = \sum_{i=0}^{8} \binom{8}{i} x^{8-2i}.\end{aligned}$$

Note that the power of x starts with 8 and descends by 2 from term to term. When $i = 4$ you procure x^0 with the corresponding **binomial coefficient** $\binom{8}{4}$.

5.4 Exercises

In problems 1—16, **simplify** the following expressions:

1. $\frac{9!}{3!6!}$.
2. $\frac{10!}{5!5!}$.
3. $\binom{6}{4}\binom{4}{2}$.
4. $\binom{8}{2}\binom{6}{4}$.
5. $\frac{7}{3}\binom{6}{2}$.
6. $\frac{5}{8}\binom{8}{3}$.
7. $\frac{(k+2)!}{k!}$, $k \in \mathbb{N}$.
8. $\frac{(k+3)!}{k!}$, $k \in \mathbb{N}$.

9. $\frac{(k+n)!}{k!}$, $k, n \in \mathbb{N}$.
10. $k \cdot [k! + (k-1)!]$, $k \in \mathbb{N}$.
11. $\frac{8!}{4!}$.
12. $\frac{12!}{6!}$.
13. $\frac{(2k)!}{k!}$, $k \in \mathbb{N}$.
14. $\dfrac{\binom{n}{k+1}}{\binom{n}{k}}$, $k, n \in \mathbb{N}$.
15. $\binom{n}{k-1} + \binom{n-1}{k} + \binom{n-1}{k-1}$.
16. $\binom{n+1}{k+1} + \binom{n}{k-1} + \binom{n-1}{k} + \binom{n-1}{k-1}$.

In problems 17–26, using the **binomial expansion** determine:

17. $(x+y+z)^2$.
18. $(x+y+z)^3$.
19. $(a^3+b^3)^6$.
20. $(a+b^2)^8$.
21. $(x+x^{-2})^{12}$.
22. $(x-y)^n$.
23. The binomial coefficient of x^3 in $(x+x^{-2})^{18}$.
24. The binomial coefficient of x^4 in $(x-x^{-3})^{16}$.
25. The binomial coefficient of x^0 in $(x+x^{-1})^{2n}$.
26. The binomial coefficient of x^3 in $(x+x^{-1})^{2n+1}$.

In problems 27–34, using Eq. (5.1) **prove** the following expressions:

27. $\binom{2k}{k}$ is even, $k \in \mathbb{N}$.
28. $\binom{2k}{2} = 2\binom{k}{2} + k^2$, $k \in \mathbb{N}$.
29. $\binom{3k}{3} = 3\binom{k}{3} + 6k\binom{k}{2} + k^3$, $k \in \mathbb{N}$.
30. $\binom{n}{k} = \frac{n}{k}\binom{n-1}{k-1}$, $k, n \in \mathbb{N}$.
31. $\binom{n}{k} = \frac{n}{n-k}\binom{n-1}{k}$, $k, n \in \mathbb{N}$.

32. $\binom{n}{k} + \binom{n}{k+1} = \binom{n+1}{k+1}$, $k, n \in \mathbb{N}$.

33. $\binom{n}{k}\binom{k}{j} = \binom{n}{j}\binom{n-j}{k-j}$, $j \leq k \leq n$.

34. $\binom{n}{n-2} + \binom{n+1}{n-1} = n^2$, $n \geq 2$.

In problems 35–40, prove the following expressions by **induction**:

35. $\sum_{i=0}^{n-1} \binom{i+1}{i} = \binom{n+1}{n-1}$, $n \geq 2$.

36. $\sum_{i=0}^{n-1} \binom{i+2}{i} = \binom{n+2}{n-1}$, $n \geq 2$.

37. $\sum_{i=0}^{n-1} \binom{i+3}{i} = \binom{n+3}{n-1}$, $n \geq 2$.

38. $\sum_{i=0}^{n-1} \binom{i+4}{i} = \binom{n+4}{n-1}$, $n \geq 2$.

39. $\sum_{i=0}^{n-1} \binom{i+k}{i} = \binom{n+k}{n-1}$, $n \geq 2, k \in \mathbb{N}$.

40. $\sum_{i=0}^{n} \binom{n}{i} = 2^n$, $n \in \mathbb{N}$.

41. Using Exercise 40, prove:

$$\sum_{i=0}^{n} \binom{2n+1}{i} = 4^n, \quad n \in \mathbb{N}.$$

42. Using Exercise 40, for even integers $n = 2, 4, 6, \ldots$, prove:

$$\sum_{i=0}^{n} [i+1]\binom{n}{i} = [n+2]2^{n-1}, \quad n \in \mathbb{N}.$$

43. Using Exercise 40, for odd integers $n = 3, 5, 7, \ldots$, prove:

$$\sum_{i=0}^{n} [i+1]\binom{n}{i} = [n+2]2^{n-1}, \quad n \in \mathbb{N}.$$

44. Using Exercise 40, prove:

$$\sum_{i=0}^{n} \binom{n}{i} 2^i = 3^n, \quad n \in \mathbb{N}.$$

Chapter 6

Recursive Relations

This chapter's aims are to formulate sequences as recursive relations while introducing the first order recursive sequences. Recall various sorts of sequences that you examined from the preceding chapters such as:

(i) Linear sequence: 1, 5, 9, 13, 17, ...
(ii) Summation-type sequence: 4, 8, 16, 28, 44, ...
(iii) Geometric sequence: 10, 20, 40, 80, 160, ...
(iv) Product-type sequence: 2, 4, 16, 96, 768, ...

6.1 Formulating a Recursive Relation

This section's aims are to alternatively formulate analogous sequences and supplemental sequences as **recursive relations**. This will then direct you to deeper understanding of the new categories of sequences and their unique traits. You will emerge your study of a linear sequence expressed as a **recursive relation** and as an **initial value problem**. The first example will formulate a recursive sequence that recites one more than multiples of 4.

Example 6.1. Write a recursive formula for:

$$1,\ 5,\ 9,\ 13,\ 17,\ 21,\ 25,\ \ldots. \tag{6.1}$$

Solution: In (6.1), you transition from neighbor to neighbor by adding a 4 and formulate (6.1) analytically as:

$$\{4n+1\}_{n=0}^{\infty}.$$

Next, formulate (6.1) recursively. By iteration and induction you acquire:

$$\begin{aligned}
x_0 &= 1, \\
x_1 &= x_0 + 4 = 1 + 4 = 5, \\
x_2 &= x_1 + 4 = 5 + 4 = 9, \\
x_3 &= x_2 + 4 = 9 + 4 = 13, \\
x_4 &= x_3 + 4 = 13 + 4 = 17, \\
x_5 &= x_4 + 4 = 17 + 4 = 21, \\
&\vdots
\end{aligned} \tag{6.2}$$

Via (6.2), for all $n \geq 0$, you formulate the following **initial value problem:**

$$\begin{cases} x_{n+1} = x_n + 4, \\ x_0 = 1. \end{cases}$$

This is a special case of a first order linear non-homogeneous difference equation.

The successive two examples will focus on formulating **summation-type** sequences recursively. The upcoming example will be formulating perfect squares recursively.

Example 6.2. Write a recursive formula for:

$$1,\ 4,\ 9,\ 16,\ 25,\ 36,\ 49,\ \ldots. \tag{6.3}$$

Solution: (6.3) lists perfect squares starting with 1 as:

$$\{n^2\}_{n=1}^{\infty}.$$

Next, formulate (6.3) recursively. By iteration you acquire:

$$\begin{aligned}
x_0 &= 1, \\
x_1 &= x_0 + 3 = 1 + (2 \cdot 0 + 3) = 4, \\
x_2 &= x_1 + 5 = 4 + (2 \cdot 1 + 3) = 9, \\
x_3 &= x_2 + 7 = 9 + (2 \cdot 2 + 3) = 16, \\
x_4 &= x_3 + 9 = 16 + (2 \cdot 3 + 3) = 25, \\
x_5 &= x_4 + 11 = 25 + (2 \cdot 4 + 3) = 36, \\
&\vdots
\end{aligned} \tag{6.4}$$

Via (6.4), for all $n \geq 0$, you formulate the following **initial value problem**:

$$\begin{cases} x_{n+1} = x_n + (2 \cdot n + 3), \\ \quad x_0 = 1. \end{cases}$$

This is special case of a first order linear non-autonomous difference equation in the **additive form**.

The successive example will recursively formulate a summation that adds consecutive even integers starting with 2.

Example 6.3. Write a recursive formula for:

$$2, \ 6, \ 12, \ 20, \ 30, \ 42, \ 56, \ \ldots. \tag{6.5}$$

Solution: Equation (6.5) depicts the sum of the consecutive positive even integers starting with 2 in the corresponding sigma notation:

$$\sum_{i=1}^{n} 2i.$$

Next, formulate (6.5) recursively. By iteration you obtain:

$$\begin{aligned} x_0 &= 2, \\ x_1 &= x_0 + 4 = 2 + 2 \cdot 2 = 6, \\ x_2 &= x_1 + 6 = 6 + 2 \cdot 3 = 12, \\ x_3 &= x_2 + 8 = 12 + 2 \cdot 4 = 20, \\ x_4 &= x_3 + 10 = 20 + 2 \cdot 5 = 30, \\ x_5 &= x_4 + 12 = 30 + 2 \cdot 6 = 42, \\ &\vdots \end{aligned} \tag{6.6}$$

Via (6.6), for all $n \geq 0$, you formulate the following **initial value problem**:

$$\begin{cases} x_{n+1} = x_n + 2(n + 2), \\ \quad x_0 = 2. \end{cases}$$

This is another special case of a first order linear non-autonomous difference equation in the **additive form**.

The next two examples will shift your focus on formulating **geometric sequences** and **product-type sequences** recursively and examine difference equations in the **multiplicative form**. The upcoming example will formulate a **geometric sequence** recursively.

Example 6.4. Write a recursive formula for:

$$5,\ 20,\ 80,\ 320,\ 1{,}280,\ 5{,}120,\ 20{,}480,\ \ldots. \tag{6.7}$$

Solution: In (6.7), you transition from neighbor to neighbor by multiplying by 4 and obtain:

$$\{5 \cdot 4^n\}_{n=0}^{\infty}.$$

Next, formulate (6.7) recursively. By iteration and induction you obtain:

$$\begin{aligned} x_0 &= 5, \\ x_1 &= 4 \cdot x_0 = 4 \cdot 5 = 20, \\ x_2 &= 4 \cdot x_1 = 4 \cdot 20 = 80, \\ x_3 &= 4 \cdot x_2 = 4 \cdot 80 = 320, \\ x_4 &= 4 \cdot x_3 = 4 \cdot 320 = 1{,}280, \\ x_5 &= 4 \cdot x_4 = 4 \cdot 1{,}280 = 5{,}120, \\ &\vdots \end{aligned} \tag{6.8}$$

Via (6.8), for all $n \geq 0$, you formulate the associated **initial value problem**:

$$\begin{cases} x_{n+1} = 4x_n, \\ x_0 = 5. \end{cases}$$

This is a special case of a first order linear homogeneous difference equation.

The sequential example resembles the **factorial pattern** recursively.

Example 6.5. Write a recursive formula for:

$$1,\ 1,\ 2,\ 6,\ 24,\ 120,\ 720,\ \ldots. \tag{6.9}$$

Solution: Equation (6.9) is the product of consecutive positive integers starting with 1 and formulate (6.9) recursively as:

$$\begin{aligned}
x_0 &= 1,\\
x_1 &= 1 \cdot x_0 = 1 \cdot 1 = 1,\\
x_2 &= 2 \cdot x_1 = 2 \cdot 1 = 2,\\
x_3 &= 3 \cdot x_2 = 3 \cdot 2 = 6,\\
x_4 &= 4 \cdot x_3 = 4 \cdot 6 = 24,\\
x_5 &= 5 \cdot x_4 = 5 \cdot 24 = 120,\\
&\vdots
\end{aligned} \tag{6.10}$$

Via (6.10), for all $n \geq 0$, you formulate the related **initial value problem**:

$$\begin{cases} x_{n+1} = (n+1) \cdot x_n, \\ \quad x_0 = 1. \end{cases}$$

This example describes the **factorial pattern** and is a special case of a first order linear non-autonomous difference equation in the **multiplicative form**.

The succeeding example depicts the product of two consecutive positive integers by multiplying and dividing when shifting from neighbor to neighbor.

Example 6.6. Write a recursive formula for:

$$2 \cdot 3,\ 3 \cdot 4,\ 4 \cdot 5,\ 5 \cdot 6,\ 6 \cdot 7,\ 7 \cdot 8,\ \ldots. \tag{6.11}$$

Solution: Formulate (6.11) with the corresponding iterative pattern:

$$\begin{aligned}
x_0 &= 2 \cdot 3,\\
x_1 &= 3 \cdot 4 = \frac{[2 \cdot 3] \cdot 4}{2} = \left(\frac{4}{2}\right) x_0,\\
x_2 &= 4 \cdot 5 = \frac{[3 \cdot 4] \cdot 5}{3} = \left(\frac{5}{3}\right) x_1,\\
x_3 &= 5 \cdot 6 = \frac{[4 \cdot 5] \cdot 6}{4} = \left(\frac{6}{4}\right) x_2,
\end{aligned}$$

$$x_4 = 6 \cdot 7 = \frac{[5 \cdot 6] \cdot 7}{5} = \left(\frac{7}{5}\right) x_3,$$

$$\vdots$$

Thus, for all $n \geq 0$ you obtain the corresponding **initial value problem** presenting (6.11):

$$\begin{cases} x_{n+1} = \left(\frac{n+4}{n+2}\right) x_n, \\ x_0 = 2 \cdot 3. \end{cases}$$

The next aim is to solve recursive sequences explicitly by inductively obtaining a formula and solving an **initial value problem.**

6.2 Obtaining an Explicit Solution

This section's goal is to inductively determine and formulate an explicit solution to assorted recursive sequences such as:

(i) Linear homogeneous: $x_{n+1} = 3x_n, \quad n = 0, 1, \ldots.$
(ii) Linear non-homogeneous: $x_{n+1} = 3x_n + 8, \quad n = 0, 1, \ldots.$
(iii) Linear non-autonomous: $x_{n+1} = a_n x_n, \quad n = 0, 1, \ldots.$
(iv) Linear non-autonomous: $x_{n+1} = a_n x_n + b_n, \quad n = 0, 1, \ldots.$

You will commence with formulating an explicit solution to the following linear homogeneous recursive relation:

$$x_{n+1} = a x_n, \quad n = 0, 1, \ldots, \tag{6.12}$$

where $a \neq 0$. By iterations and induction you get:

$$\begin{aligned} & x_0, \\ & x_1 = a \cdot x_0, \\ & x_2 = a \cdot x_1 = a \cdot [a x_0] = a^2 x_0, \\ & x_3 = a \cdot x_2 = a \cdot \left[a^2 x_0\right] = a^3 x_0, \\ & x_4 = a \cdot x_3 = a \cdot \left[a^3 x_0\right] = a^4 x_0, \\ & x_5 = a \cdot x_4 = a \cdot \left[a^4 x_0\right] = a^5 x_0, \\ & \vdots \end{aligned} \tag{6.13}$$

Via (6.18), for all $n \in \mathbb{N}$ you procure the corresponding solution to Eq. (6.12):

$$x_n = a^n x_0. \tag{6.14}$$

Note that (6.14) is a geometric sequence. The consequent example will apply Eq. (6.12) in obtaining an explicit solution to an **initial value problem**.

Example 6.7. Solve the given **initial value problem**:

$$\begin{cases} 4x_{n+1} - 3x_n = 0, \quad n = 0, 1, \ldots, \\ x_0 = \frac{9}{16}, \end{cases}$$

and verify that the solution is correct.

Solution: Via Eq. (6.14) you obtain the following solution:

$$x_n = x_0 \left(\frac{3}{4}\right)^n = \frac{9}{16}\left(\frac{3}{4}\right)^n = \left(\frac{3}{4}\right)^2 \left(\frac{3}{4}\right)^n = \left(\frac{3}{4}\right)^{n+2}. \tag{6.15}$$

Then via (6.15) you acquire:

$$x_{n+1} = \left(\frac{3}{4}\right)^{n+3}. \tag{6.16}$$

Now via (6.15) and (6.16) you obtain:

$$4x_{n+1} - 3x_n = 4\left(\frac{3}{4}\right)^{n+3} - 3\left(\frac{3}{4}\right)^{n+2} = \frac{3^{n+3}}{4^{n+2}} - \frac{3^{n+3}}{4^{n+2}} = 0.$$

The result follows.

The next aim is to obtain an explicit solution to the corresponding linear non-homogeneous recursive relation:

$$x_{n+1} = ax_n + b, \quad n = 0, 1, \ldots, \tag{6.17}$$

where $a \neq 0, 1$ and $b \neq 0$. By iterations and induction you procure:

$$\begin{aligned}
&x_0,\\
&x_1 = ax_0 + b,\\
&x_2 = ax_1 + b = a \cdot [ax_0 + b] + b = a^2 x_0 + a \cdot b + b,\\
&x_3 = ax_2 + b = a \cdot \left[a^2 x_0 + a \cdot b + b\right] + b = a^3 x_0 + a^2 \cdot b + a \cdot b + b,\\
&\vdots
\end{aligned} \tag{6.18}$$

Thus via (6.18), for all $n \in \mathbb{N}$ you obtain the following solution to Eq. (6.17):

$$x_n = a^n x_0 + b\left[\sum_{i=0}^{n-1} a^i\right] = a^n x_0 + b\left[\frac{1-a^n}{1-a}\right]. \tag{6.19}$$

(6.19) is obtained by induction to a first order linear non-homogeneous difference equation. You can also reformulate (6.19) as:

$$x_n = a^n\left[x_0 - \frac{b}{1-a}\right] + \frac{b}{1-a}. \tag{6.20}$$

The consequent example will present the application of (6.20) in solving an **initial value problem** of Eq. (6.17).

Example 6.8. Solve the designated **initial value problem**:

$$\begin{cases} 4x_{n+1} - x_n = 9. & n = 0, 1, \ldots, \\ \qquad x_0 = 7, \end{cases}$$

and verify that the solution is correct.

Solution: From (6.20), you obtain:

$$x_n = \left(\frac{1}{4}\right)^n [7-3]+3 = 4\left(\frac{1}{4}\right)^n + 3 = \left(\frac{1}{4}\right)^{n-1} + 3. \quad (6.21)$$

Then via (6.15), you acquire:

$$x_{n+1} = \left(\frac{1}{4}\right)^n + 3. \quad (6.22)$$

Now via (6.21) and (6.22), you obtain:

$$\begin{aligned} 4x_{n+1} - x_n &= 4\left[\left(\frac{1}{4}\right)^n + 3\right] - \left[\left(\frac{1}{4}\right)^{n-1} + 3\right] \\ &= \frac{1}{4^{n-1}} + 12 - \frac{1}{4^{n-1}} - 3 = 9. \end{aligned}$$

The result follows.

The upcoming sequence of examples will focus on inductively formulating a general solution to specific recursive relations.

Example 6.9. Formulate an **explicit solution** to the following recursive relation:

$$x_{n+1} = x_n + b, \quad n = 0, 1, \ldots, \quad (6.23)$$

where $b \neq 0$.

Solution: By iterations and induction you procure:

$$\begin{aligned} &x_0, \\ &x_1 = x_0 + b, \\ &x_2 = x_1 + b = [x_0 + b] + b = x_0 + 2b, \\ &x_3 = x_2 + b = [x_0 + 2b] + b = x_0 + 3b, \\ &x_4 = x_3 + b = [x_0 + 3b] + b = x_0 + 4b, \\ &x_5 = x_4 + b = [x_0 + 4b] + b = x_0 + 5b, \\ &\vdots \end{aligned} \quad (6.24)$$

Thus, via (6.24), for all $n \geq 0$ you obtain the following solution to Eq. (6.23):

$$x_n = x_0 + nb. \quad (6.25)$$

Verifying (6.25) by induction will be left as an end of chapter exercise.

Example 6.10. Formulate an **explicit solution** to the following recursive relation:

$$x_{n+1} = -x_n + b, \quad n = 0, 1, \ldots, \tag{6.26}$$

where $b \neq 0$.

Solution: By iterations and induction you procure:

$$\begin{aligned} &x_0, \\ &x_1 = x_0 + b, \\ &x_2 = -x_1 + b = -[-x_0 + b] + b = x_0, \\ &x_3 = -x_2 + b = -[x_0] + b = x_0 + b, \\ &x_4 = -x_3 + b = -[-x_0 + b] + b = x_0, \\ &\vdots \end{aligned} \tag{6.27}$$

Via (6.27), for all $n \geq 0$ you formulate the following solution to Eq. (6.26):

$$x_n = \begin{cases} x_0 & \text{if } n = 0, 2, 4, 6, \ldots, \\ x_0 + b & \text{if } n = 1, 3, 5, 7 \ldots. \end{cases}$$

This is an example of a period-2 cycle and a piecewise solution to a first order linear non-homogeneous recursive sequence.

Example (6.10) will guide to periodic recursive relations in Chapter 7. The next section will focus on formulating an explicit solution to a first order linear non-autonomous difference equation.

6.3 Non-Autonomous Recursive Sequences

This section's aim is to inductively formulate an explicit solution to specific non-autonomous recursive sequences either in the additive form, multiplicative form or as a combination of the two forms. You will examine the following three categories of first order non-autonomous linear recursive sequences:

(i) Linear non-autonomous: $x_{n+1} = x_n + b_n, \quad n = 0, 1, \ldots,$
(ii) Linear non-autonomous: $x_{n+1} = a_n x_n, \quad n = 0, 1, \ldots,$
(iii) Linear non-autonomous: $x_{n+1} = a_n x_n + b_n, \quad n = 0, 1, \ldots,$

where $\{a_n\}_{n=0}^{\infty}$ and $\{b_n\}_{n=0}^{\infty}$ are sequences of real numbers.

In (i), the recursive relation is in the additive form, where for all $n \geq 0$ each term of the sequence $\{b_n\}_{n=0}^{\infty}$ is added to x_n during each iteration.

In (ii), the recursive relation is in the multiplicative form, where for all $n \geq 0$ each term of the sequence $\{a_n\}_{n=0}^{\infty}$ is multiplied by x_n during each iteration.

In (iii), the recursive relation is a combination of two forms, where for all $n \geq 0$, b_n is added to the product $a_n x_n$.

The next three Sections 6.3.1–6.3.3 will examine the three different cases of the corresponding non-autonomous recursive relation:

$$x_{n+1} = a_n x_n + b_n, \quad n = 0, 1, \ldots, \tag{6.28}$$

where $\{a_n\}_{n=0}^{\infty}$ and $\{b_n\}_{n=0}^{\infty}$ are sequences of real numbers. The upcoming Section 6.3.1, will focus on a special case of Eq. (6.28) in the **additive form**.

6.3.1 *Additive Form of Eq. (6.28)*

In this section, your goals are to formulate an explicit solution to the corresponding non-autonomous homogeneous linear recursive sequence in the **additive form**:

$$x_{n+1} = x_n + b_n \quad n = 0, 1, \ldots, \tag{6.29}$$

where $\{b_n\}_{n=0}^{\infty}$ is a sequence of real numbers. By induction and iterations you acquire:

$$\begin{aligned}
&x_0, \\
&x_1 = x_0 + b_0, \\
&x_2 = x_1 + b_1 = x_0 + [b_0 + b_1], \\
&x_3 = x_2 + b_2 = x_0 + [b_0 + b_1 + b_2], \\
&x_4 = x_3 + b_3 = x_0 + [b_0 + b_1 + b_2 + b_3], \\
&x_5 = x_4 + b_4 = x_0 + [b_0 + b_1 + b_2 + b_3 + b_4], \\
&x_6 = x_5 + b_5 = x_0 + [b_0 + b_1 + b_2 + b_3 + b_4 + b_5], \\
&\vdots
\end{aligned} \tag{6.30}$$

Via (6.30), for all $n \in \mathbb{N}$ you formulate the corresponding solution of Eq. (6.29):

$$x_n = x_0 + [b_0 + b_1 + \cdots + b_{n-1}] = x_0 + \sum_{i=0}^{n-1} b_i. \tag{6.31}$$

The succeeding four examples will present the use of (6.31) while inductively formulating the solution to (6.29). The consequent example adds consecutive positive even integers and applies (3.3).

Example 6.11. Formulate an **explicit solution** of the following recursive relation:

$$x_{n+1} = x_n + (2n + 2), \quad n = 0, 1, \ldots. \tag{6.32}$$

Solution: By iterations and induction you get:

$$\begin{aligned}
&x_0,\\
&x_1 = x_0 + 2,\\
&x_2 = x_1 + 4 = x_0 + [2 + 4],\\
&x_3 = x_2 + 6 = x_0 + [2 + 4 + 6],\\
&x_4 = x_3 + 8 = x_0 + [2 + 4 + 6 + 8],\\
&x_5 = x_4 + 10 = x_0 + [2 + 4 + 6 + 8 + 10],\\
&x_6 = x_5 + 12 = x_0 + [2 + 4 + 6 + 8 + 10 + 12],\\
&\vdots
\end{aligned} \tag{6.33}$$

Via (6.33) and (3.3), for all $n \in \mathbb{N}$ you formulate the corresponding solution to Eq. (6.32):

$$x_n = x_0 + \sum_{i=1}^{n} 2i = x_0 + n[n + 1].$$

The upcoming example adds consecutive geometric terms and applies (3.6).

Example 6.12. Formulate an **explicit solution** of the following recursive relation:

$$x_{n+1} = x_n + 3^n, \quad n = 0, 1, \ldots. \tag{6.34}$$

Solution: By iterations and induction you get:

$$\begin{aligned}
&x_0, \\
&x_1 = x_0 + 3^0, \\
&x_2 = x_1 + 3^1 = x_0 + \left[3^0 + 3^1\right], \\
&x_3 = x_2 + 3^2 = x_0 + \left[3^0 + 3^1 + 3^2\right], \\
&x_4 = x_3 + 3^3 = x_0 + \left[3^0 + 3^1 + 3^2 + 3^3\right], \\
&x_5 = x_4 + 3^4 = x_0 + \left[3^0 + 3^1 + 3^2 + 3^3 + 3^4\right], \\
&x_6 = x_5 + 3^5 = x_0 + \left[3^0 + 3^1 + 3^2 + 3^3 + 3^4 + 3^5\right], \\
&\vdots
\end{aligned} \tag{6.35}$$

Via (6.35) and (3.6), for all $n \in \mathbb{N}$ you formulate the corresponding solution to Eq. (6.34):

$$x_n = x_0 + \sum_{i=0}^{n-1} 3^i = x_0 + \left[\frac{3^n - 1}{2}\right].$$

The upcoming two examples will formulate a piecewise and an alternating solution to specific recursive relations.

Example 6.13. Formulate an **explicit piecewise solution** of the following **initial value problem**:

$$\begin{cases} x_{n+1} = -x_n + (n+1), \quad n = 0, 1, 2, \ldots, \\ x_0 = 0 \end{cases} \tag{6.36}$$

Solution: By iterations and induction you acquire:

$$\begin{aligned}
x_0 &= 0, \\
x_1 &= -\left[x_0\right] + 1 = -\left[0\right] + 1 = 1, \\
x_2 &= -\left[x_1\right] + 2 = -\left[1\right] + 2 = 1, \\
x_3 &= -\left[x_2\right] + 3 = -\left[1\right] + 3 = 2, \\
x_4 &= -\left[x_3\right] + 4 = -\left[2\right] + 4 = 2, \\
x_5 &= -\left[x_4\right] + 5 = -\left[2\right] + 5 = 3, \\
x_6 &= -\left[x_5\right] + 6 = -\left[3\right] + 6 = 3, \\
x_7 &= -\left[x_6\right] + 7 = -\left[3\right] + 7 = 4, \\
&\vdots
\end{aligned} \tag{6.37}$$

Via (6.37), for all $n \geq 0$ you formulate the corresponding piecewise solution to Eq. (6.36):

$$x_n = \begin{cases} \frac{n}{2} & \text{if } n = 0, 2, 4, 6, \ldots, \\ \frac{n+1}{2} & \text{if } n = 1, 3, 5, 7, \ldots. \end{cases}$$

The consequent example will formulate an alternating solution as a piecewise geometric sequence.

Example 6.14. Formulate an **alternating explicit solution** of the following recursive relation:

$$x_{n+1} = -x_n + 2^n, \quad n = 0, 1, \ \ldots. \tag{6.38}$$

Solution: By iterations and induction you acquire:

$$\begin{aligned} & x_0, \\ & x_1 = -x_0 + [1], \\ & x_2 = -x_1 + 2 = x_0 - [1 - 2], \\ & x_3 = -x_2 + 2^2 = -x_0 + [1 - 2 + 2^2], \\ & x_4 = -x_3 + 2^3 = x_0 - [1 - 2 + 2^2 - 2^3], \\ & x_5 = -x_4 + 2^4 = -x_0 + [1 - 2 + 2^2 - 2^3 + 2^4], \\ & x_6 = -x_5 + 2^5 = x_0 - [1 - 2 + 2^2 - 2^3 + 2^4 - 2^5], \\ & x_7 = -x_6 + 2^6 = -x_0 + [1 - 2 + 2^2 - 2^3 + 2^4 - 2^5 + 2^6], \\ & \vdots \end{aligned} \tag{6.39}$$

Via (6.39) and (3.6), for all $n \in \mathbb{N}$ you formulate the corresponding piecewise solution to Eq. (6.38):

$$x_n = \begin{cases} x_0 - \sum_{i=0}^{n-1} (-2)^i & \text{if } n = 0, 2, 4, 6, \ldots, \\ = x_0 - \left[\frac{1-(-2)^n}{3}\right] & \\ -x_0 + \sum_{i=0}^{n-1} (-2)^i & \text{if } n = 1, 3, 5, 7, \ldots. \\ = -x_0 + \left[\frac{1-(-2)^n}{3}\right] & \end{cases} \tag{6.40}$$

For all $n \in \mathbb{N}$ you reformulate (6.40) as the corresponding alternating solution to Eq. (6.38):

$$x_n = (-1)^n x_0 + (-1)^{n+1} \sum_{i=0}^{n-1} (-1)^i 2^i = (-1)^n x_0 + (-1)^{n+1} \left[\frac{1 - (-2)^n}{3}\right].$$

The upcoming Section 6.3.2 will transition to examining a special case of Eq. (6.28) in the **multiplicative form**.

6.3.2 *Multiplicative Form of Eq. (6.28)*

This section's objectives are to formulate an explicit solution of the corresponding non-autonomous homogeneous linear recursive sequence in the **multiplicative form**:

$$x_{n+1} = a_n x_n, \quad n = 0, 1, \ldots, \tag{6.41}$$

where $\{a_n\}_{n=0}^{\infty}$ is a sequence of real numbers. By iterations and induction you acquire:

$$\begin{aligned}
&x_0, \\
&x_1 = a_0 x_0, \\
&x_2 = a_1 x_1 = [a_0 a_1]\, x_0, \\
&x_3 = a_2 x_2 = [a_0 a_1 a_2]\, x_0, \\
&x_4 = a_3 x_3 = [a_0 a_1 a_2 a_3]\, x_0, \\
&x_5 = a_4 x_4 = [a_0 a_1 a_2 a_3 a_4]\, x_0, \\
&x_6 = a_5 x_5 = [a_0 a_1 a_2 a_3 a_4 a_5]\, x_0, \\
&\vdots
\end{aligned} \tag{6.42}$$

Via (6.42), for all $n \in \mathbb{N}$ you formulate the cognate solution of Eq. (6.41):

$$x_n = [a_0 \cdot a_1 \cdot \quad \cdots \quad \cdot a_{n-1}]\, x_0 = \left[\prod_{i=0}^{n-1} a_i\right] x_0. \tag{6.43}$$

The upcoming examples will present the use of (6.43) while inductively formulating the solution to (6.41). The consequent example combines a geometric pattern together with adding consecutive positive integers and applies (3.3).

Example 6.15. Formulate the **explicit solution** of the following recursive relation:

$$x_{n+1} = 4^{n+1} x_n, \quad n = 0, 1, \ldots, \tag{6.44}$$

where $x_0 \neq 0$.

Solution: By iteration you acquire:

$$
\begin{aligned}
&x_0,\\
&x_1 = 4x_0,\\
&x_2 = 4^2x_1 = 4^2\,[4x_0] = 4^{[1+2]}x_0,\\
&x_3 = 4^3x_2 = 4^3\left[4^{[1+2]}x_0\right] = 4^{[1+2+3]}x_0,\\
&x_4 = 4^4x_3 = 4^4\left[4^{[1+2+3]}x_0\right] = 4^{[1+2+3+4]}x_0,\\
&x_5 = 4^5x_4 = 4^5\left[4^{[1+2+3+4]}x_0\right] = 4^{[1+2+3+4+5]}x_0,\\
&x_6 = 4^6x_5 = 4^6\left[4^{[1+2+3+4+5]}x_0\right] = 4^{[1+2+3+4+5+6]}x_0,\\
&\vdots
\end{aligned}
\tag{6.45}
$$

Via (6.45) and (3.3), for all $n \in \mathbb{N}$ you formulate the related solution to Eq. (6.44):

$$x_n = \left[\prod_{i=1}^{n} 4^i\right] x_0 = 4^{\left[\sum_{i=1}^{n} i\right]}x_0 = 4^{\left[\frac{n(n+1)}{2}\right]}x_0 = 2^{n[n+1]}x_0.$$

Note that Example (6.15) can be analyzed in terms of two patterns; a geometric pattern in blue and a summation-type pattern in red. The consequent example formulates the solution as a product of two patterns; the geometric pattern and the factorial pattern.

Example 6.16. Formulate the **explicit solution** of the following recursive relation:

$$x_{n+1} = 2(n+1)x_n, \quad n = 0, 1, \ldots, \tag{6.46}$$

where $x_0 \neq 0$.

Solution: By iteration and induction you obtain:

$$
\begin{aligned}
&x_0,\\
&x_1 = 2\cdot 1x_0,\\
&x_2 = 2\cdot 2x_1 = 2^2\cdot[2\cdot 1]\,x_0,\\
&x_3 = 2\cdot 3x_2 = 2^3\cdot[3\cdot 2\cdot 1]\,x_0,\\
&x_4 = 2\cdot 4x_3 = 2^4\cdot[4\cdot 3\cdot 2\cdot 1]\,x_0,\\
&x_5 = 2\cdot 5x_4 = 2^5\cdot[5\cdot 4\cdot 3\cdot 2\cdot 1]\,x_0,\\
&x_6 = 2\cdot 6x_4 = 2^6\cdot[6\cdot 5\cdot 4\cdot 3\cdot 2\cdot 1]\,x_0,\\
&\vdots
\end{aligned}
\tag{6.47}
$$

Via (6.47), for all $n \in \mathbb{N}$ you formulate the related solution to Eq. (6.46):

$$x_n = 2^n \left[\prod_{i=1}^{n} i\right] x_0 = 2^n n!\, x_0.$$

The succeeding Section 6.3.3 will focus on a special case of Eq. (6.28) in the **additive and multiplicative form**.

6.3.3 *Additive and Multiplicative Form of Eq. (6.28)*

This section's aims are to formulate an explicit solution of the non-autonomous non-homogeneous linear recursive sequence in the corresponding **additive and multiplicative form**:

$$x_{n+1} = a_n x_n + b_n, \quad n = 0, 1, \ldots, \tag{6.48}$$

where $\{a_n\}_{n=0}^{\infty}$ and $\{b_n\}_{n=0}^{\infty}$ are sequences of real numbers. By induction and iterations you acquire:

$$\begin{aligned}
&x_0,\\
&x_1 = a_0 x_0 + b_0,\\
&x_2 = a_1 x_1 + b_1 = a_0 a_1 x_0 + a_1 b_0 + b_1,\\
&x_3 = a_2 x_2 + b_2 = a_0 a_1 a_2 x_0 + a_1 a_2 b_0 + a_2 b_1 + b_2,\\
&x_4 = a_3 x_3 + b_3 = a_0 a_1 a_2 a_3 x_0 + a_1 a_2 a_3 b_0 + a_2 a_3 b_1 + a_3 b_2 + b_3,\\
&x_5 = a_4 x_4 + b_4 = a_0 a_1 a_2 a_3 a_4 x_0 + a_1 a_2 a_3 a_4 b_0 + a_2 a_3 a_4 b_1\\
&\qquad + a_3 a_4 b_2 + a_4 b_3 + b_4,\\
&\vdots
\end{aligned} \tag{6.49}$$

Via (6.49), for all $n \in \mathbb{N}$ you formulate the cognate solution to Eq. (6.48):

$$x_n = \prod_{i=0}^{n-1} a_i x_0 + \sum_{j=0}^{n-1} \left[\prod_{i=j+1}^{n-1} a_i b_j\right]. \tag{6.50}$$

Equation (6.50) is also an example of a **convolution**. The succeeding example renders regrouping of different patterns while formulating the solution of a recursive relation in the form (6.48).

Example 6.17. Formulate the **explicit solution** of the following the **initial value problem:**

$$\begin{cases} x_{n+1} = 2x_n + (n+2), & n = 0, 1, \ldots, \\ x_0 = 1 \end{cases} \tag{6.51}$$

Solution: By iteration and induction you obtain:

$$\begin{aligned} x_0 &= 1, \\ x_1 &= 2x_0 + 2 = 2 + 2, \\ x_2 &= 2x_1 + 3 = 2 \cdot [2 + 2] + 3, \\ &= 2^2 + \left[2 \cdot 2^1 + 3 \cdot 2^0\right], \\ x_3 &= 2x_2 + 4 = 2 \cdot [2^2 + 2 \cdot 2^1 + 3 \cdot 2^0] + 4, \\ &= 2^3 + \left[2 \cdot 2^2 + 3 \cdot 2^1 + 4 \cdot 2^0\right], \\ x_4 &= 2x_3 + 5 = 2 \cdot [2^3 + 2 \cdot 2^2 + 3 \cdot 2^1 + 4 \cdot 2^0] + 5, \\ &= 2^4 + \left[2 \cdot 2^3 + 3 \cdot 2^2 + 4 \cdot 2^1 + 5 \cdot 2^0\right], \\ x_5 &= 2x_4 + 6 = 2 \cdot [2^4 + 2 \cdot 2^3 + 3 \cdot 2^2 + 4 \cdot 2^1 + 5 \cdot 2^0] + 6, \\ &= 2^5 + \left[2 \cdot 2^4 + 3 \cdot 2^3 + 4 \cdot 2^2 + 5 \cdot 2^1 + 6 \cdot 2^0\right], \\ &\vdots \end{aligned} \tag{6.52}$$

Via (6.52), for all $n \geq 2$ you formulate the cognate solution to Eq. (6.51):

$$x_n = 2^n + \left[\sum_{i=0}^{n-1} (i+2) \cdot 2^{(n-1)-i}\right].$$

6.4 Exercises

In problems 1–8, write a **recursive formula** (as an initial value problem) of each sequence:

1. 2, 7, 12, 17, 22, 27, 32, ...
2. 7, 19, 31, 43, 55, 67, 79, ...
3. 5, 7, 11, 17, 25, 35, 47, ...

4. 4, 7, 13, 22, 34, 49, 67, ...
5. 3, 7, 15, 27, 43, 63, 87, ...
6. 2, 3, 6, 11, 18, 27, 38, ...
7. 5, 8, 17, 32, 53, 80, 113, ...
8. 1, 2, 7, 16, 29, 46, 67, ...

In problems 9–18, write a **recursive formula** (as an initial value problem) of each sequence:

9. 4, 12, 36, 108, 324, 972, ...
10. 9, 18, 36, 72, 144, 288, ...
11. 54, 36, 24, 16, $\frac{32}{3}$, $\frac{64}{9}$, ...
12. 32, 24, 18, $\frac{27}{2}$, $\frac{81}{8}$, $\frac{343}{32}$, ...
13. 1, 2, 8, 48, 384, 3,840, ...
14. 1, 3, 15, 105, 945, 10,395, ...
15. 1, 5, 45, 585, 9,945, ...
16. $1 \cdot 3$, $3 \cdot 5$, $5 \cdot 7$, $7 \cdot 9$, $9 \cdot 11$, ...
17. $2 \cdot 4$, $4 \cdot 6$, $6 \cdot 8$, $8 \cdot 10$, $10 \cdot 12$, ...
18. $1 \cdot 2 \cdot 3$, $2 \cdot 3 \cdot 4$, $3 \cdot 4 \cdot 5$, $4 \cdot 5 \cdot 6$, $5 \cdot 6 \cdot 7$, ...

In problems 19–22, write a **recursive formula** (as an initial value problem) of each summation:

19. $\sum_{k=1}^{n} (2k-1)$.
20. $\sum_{k=0}^{n} (3k+2)$.
21. $\sum_{k=1}^{n} k^2$.
22. $\sum_{k=0}^{n} \left(\frac{1}{2}\right)^k$.

In problems 23–28, show that the solution **satisfies** the given recursive sequence:

23. $x_n = \left(\frac{2}{5}\right)^{n+1}$ is a solution of $5x_{n+1} - 2x_n = 0$.

24. $x_n = \frac{9^{n+1}}{4^{n-1}}$ is a solution of $4x_{n+1} - 9x_n = 0$.

25. $x_n = \left(\frac{1}{3}\right)^n + 3$ is a solution of $3x_{n+1} - x_n = 6$.

26. $x_n = 5^{n+1} - 4$ is a solution of $x_{n+1} - 5x_n = 16$.

27. $x_n = (-4)^{n-2} + 3$ is a solution of $x_{n+1} + 4x_n = 15$.

28. $x_n = nb + 3$ is a solution of $x_{n+1} - x_n = b$.

In problems 29–42, solve the given **initial value problem** and check your answer.

29. $\begin{cases} 2x_{n+1} - x_n = 0, & n = 0, 1, \ldots. \\ x_0 = \frac{1}{8}. \end{cases}$

30. $\begin{cases} 3x_{n+1} - 2x_n = 0, & n = 0, 1, \ldots. \\ x_0 = \frac{4}{9}. \end{cases}$

31. $\begin{cases} 3x_{n+1} - 4x_n = 0, & n = 0, 1, \ldots. \\ x_0 = \frac{16}{9}. \end{cases}$

32. $\begin{cases} 5x_{n+1} - 2x_n = 6, & n = 0, 1, \ldots. \\ x_0 = 3. \end{cases}$

33. $\begin{cases} 7x_{n+1} - 3x_n = 12, & n = 0, 1, \ldots. \\ x_0 = 4. \end{cases}$

34. $\begin{cases} x_{n+1} - x_n = 4, & n = 0, 1, \ldots. \\ x_0 = -2. \end{cases}$

35. $\begin{cases} 2x_{n+1} - 2x_n = -1, & n = 0, 1, \ldots. \\ x_0 = \frac{3}{2}. \end{cases}$

36. $\begin{cases} 2x_{n+1} + 2x_n = 7, & n = 0, 1, \ldots. \\ x_0 = \frac{1}{2}. \end{cases}$

37. $\begin{cases} x_{n+1} + x_n = 7, \quad n = 0, 1, \ \ldots. \\ x_0 = -2. \end{cases}$

38. $\begin{cases} x_{n+1} - x_n = (-1)^n, \quad n = 0, 1, \ \ldots. \\ x_0 = 4. \end{cases}$

39. $\begin{cases} x_{n+1} + x_n = (-1)^{n+1}, \quad n = 0, 1, \ \ldots. \\ x_0 = 4. \end{cases}$

40. $\begin{cases} x_{n+1} - x_n = 3^{n+2}, \quad n = 0, 1, \ \ldots. \\ x_0 = 3. \end{cases}$

41. $\begin{cases} x_{n+1} - x_n = 2^{n+3}, \quad n = 0, 1, \ \ldots. \\ x_0 = 4. \end{cases}$

42. $\begin{cases} x_{n+1} - x_n = (2n + 3), \quad n = 0, 1, \ \ldots. \\ x_0 = 1. \end{cases}$

43. $\begin{cases} x_{n+1} - x_n = (n + 2)^2, \quad n = 0, 1, \ \ldots. \\ x_0 = 1. \end{cases}$

44. $\begin{cases} x_{n+1} - x_n = 2n + 2, \quad n = 0, 1, \ \ldots. \\ x_0 = 0. \end{cases}$

Chapter 7

Periodic Traits

This chapter's aims are to investigate an formulate various periodic traits of the autonomous and non-autonomous first order linear recursive relations:

(i) Linear autonomous: $x_{n+1} = -x_n, \quad n = 0, 1, \ldots,$
(ii) Linear autonomous: $x_{n+1} = -x_n + b, \quad n = 0, 1, \ldots,$
(iii) Linear non-autonomous: $x_{n+1} = \pm a_n x_n, \quad n = 0, 1, \ldots,$
(iv) Linear non-autonomous: $x_{n+1} = \pm x_n + b_n, \quad n = 0, 1, \ldots,$
(v) Linear non-autonomous: $x_{n+1} = \pm a_n x_n + b_n, \quad n = 0, 1, \ldots,$

where $\{a_n\}_{n=0}^{\infty}$ and $\{b_n\}_{n=0}^{\infty}$ are periodic sequences. In some instances you will encounter the existence of unique periodic cycles.

Definition 1. The sequence $\{x_n\}_{n=0}^{\infty}$ is periodic with **period-*p***, $(p \geq 2)$, provided that for all $n \geq 0$:

$$x_{n+p} = x_n.$$

The **minimal period** is smallest such number p. First, let's commence with graphical examples of assorted periodic cycles with various shapes such as the triangular and trapezoidal shapes. The first sketch in Figure 7.1 traces an ascending triangular-shaped period-3 cycle.

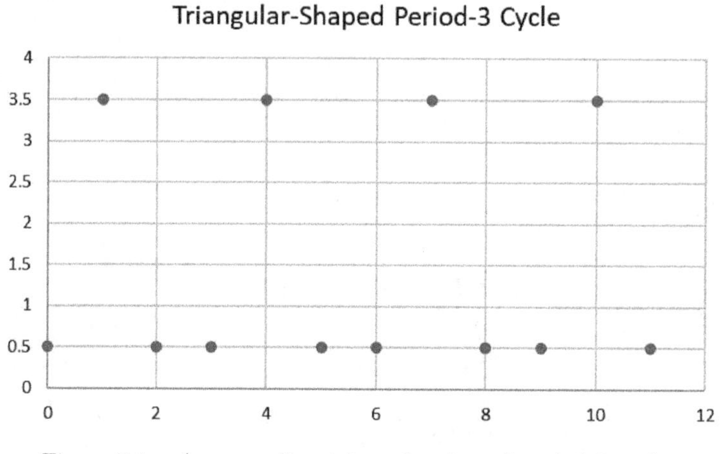

Figure 7.1: An ascending triangular-shaped period-3 cycle.

Figure 7.1 portrays an ascending triangular-shaped period-3 cycle, where for all $n \geq 0$:

$$x_{n+3} = x_n.$$

Figure 7.2 describes a trapezoidal-shaped period-4 cycle.

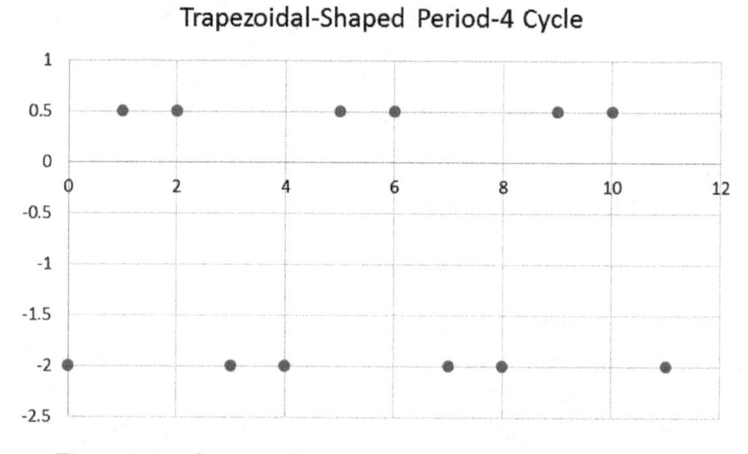

Figure 7.2: An ascending trapezoidal-shaped period-4 cycle.

Figure 7.2 resembles an ascending trapezoidal-shaped period-4 cycle, where for all $n \geq 0$:

$$x_{n+4} = x_n.$$

Figure 7.3 depicts an alternating period-6 cycle.

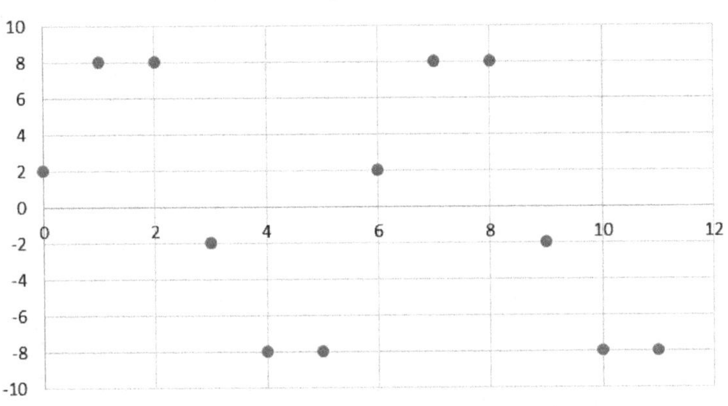

Figure 7.3: An alternating period-6 cycle.

Figure 7.3 also traces an **ascending-shaped** and a **descending-shaped** period-6 cycle, where for all $n \geq 0$:

$$x_{n+6} = x_n \quad \text{and} \quad x_{n+3} = -x_n.$$

Next, turn your focus to examining and formulating periodicity traits of the associated recursive relation:

$$x_{n+1} = a_n x_n + b_n, \qquad n = 0, 1, \ldots, \tag{7.1}$$

where $\{a_n\}_{n=0}^{\infty}$ and $\{b_n\}_{n=0}^{\infty}$ are either constant or periodic sequences. The upcoming section will focus on the autonomous special case of Eq. (7.1) where $\{a_n\}_{n=0}^{\infty}$ and $\{b_n\}_{n=0}^{\infty}$ are constant sequences.

7.1 Autonomous Recursive Sequences

We will commence our study of periodic properties of the following first order autonomous recursive relation:

$$x_{n+1} = -x_n + b, \qquad n = 0, 1, \ldots, \tag{7.2}$$

where $x_0, b \in \Re$. Eq. (7.2) consists of two cases when $b = 0$ and when $b \neq 0$. In fact, when $b = 0$, Eq. (7.2) reduces to the corresponding homogeneous linear recursive sequence:

$$x_{n+1} = -x_n, \quad n = 0, 1, \ldots, \tag{7.3}$$

where $x_0 \neq 0$. By iterations and induction you formulate the corresponding solution to Eq. (7.3):

$$x_n = (-1)^n x_0 = \begin{cases} x_0 & \text{if } n = 0, 2, 4, 6, \ldots, \\ -x_0 & \text{if } n = 1, 3, 5, 7, \ldots. \end{cases} \tag{7.4}$$

Note that Figure 1.12 in Chapter 1 is a special case of (7.4) when $x_0 = 1$.

Next, assume that $b \neq 0$. In this case, Eq. (7.3) becomes an autonomous non-homogeneous recursive sequence. By iterations and induction you formulate the associated solution to Eq. (7.2):

$$x_n = \begin{cases} x_0 & \text{if } n = 0, 2, 4, 6, \ldots, \\ -x_0 + b & \text{if } n = 1, 3, 5, 7, \ldots. \end{cases} \tag{7.5}$$

Next, transition to investigating the periodic traits of non-autonomous recursive sequences. In fact, the upcoming section will examine Eq. (7.1) in the **multiplicative form** where $\{a_n\}_{n=0}^{\infty}$ is periodic sequence.

7.2 Multiplicative Form of Eq. (7.1)

This section's goals are to examine the periodic features of Eq. (7.1) in the associated **multiplicative form**:

$$x_{n+1} = a_n x_n, \quad n = 0, 1, \ldots, \tag{7.6}$$

where $x_0 \neq 0$ and $\{a_n\}_{n=0}^{\infty}$ is a period-k sequence for ($k \geq 2$). The main objective is to examine and formulate the periodic attributes of Eq. (7.6).

First, assume that $\{a_n\}_{n=0}^{\infty}$ is a period-2 sequence and formulate the **explicit solution** of Eq. (7.6) and specify the periodicity character.

By iterations and induction you get:

$$
\begin{aligned}
&x_0, \\
&x_1 = a_0 x_0, \\
&x_2 = a_1 x_1 = [a_1 a_0]\, x_0, \\
&x_3 = a_0 x_2 = a_0 [a_1 a_0]\, x_0, \\
&x_4 = a_1 x_3 = [a_1 a_0]^2 x_0, \\
&x_5 = a_0 x_4 = a_0 [a_1 a_0]^2 x_0, \\
&x_6 = a_1 x_5 = [a_1 a_0]^3 x_0, \\
&\vdots
\end{aligned}
\tag{7.7}
$$

Let $P = a_0 a_1$. Via (7.7), for all $n \in \mathbb{N}$ you formulate the cognate solution to Eq. (7.6):

$$
\begin{cases}
x_{2n} = P^n x_0, \\
x_{2n+1} = a_0 P^n x_0.
\end{cases}
\tag{7.8}
$$

Thus via (7.8), every solution of Eq. (7.6) is periodic with:

(i) Period-2 if $P = 1$.
(ii) Period-4 if $P = -1$.

Next, you will sketch a graphical representation of a period-2 cycle when $P = 1$ and a period-4 cycle when $P = -1$. Figure 7.4 traces a **positive decreasing period-2 cycle** when $x_0 = 4$, $a_0 = 0.125$ and $a_1 = 8$.

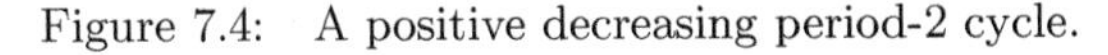

Figure 7.4: A positive decreasing period-2 cycle.

By designating distinct values of x_0, a_0 and a_1, you can acquire four combinations of periodic cycles: a positive increasing periodic cycle, a positive decreasing periodic cycle, a negative increasing periodic cycle, and a negative decreasing periodic cycle. Figure 7.5 renders an **alternating step-shaped period-4 cycle** when $x_0 = -2$, $a_0 = 1$ and $a_1 = -1$.

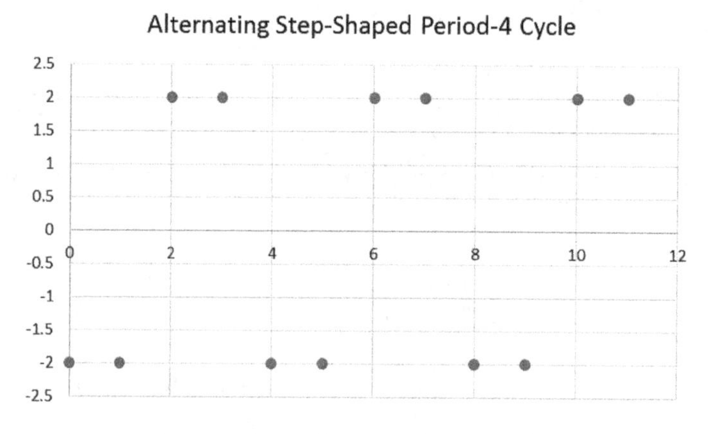

Figure 7.5: An alternating step-shaped period-4 cycle.

Figure 7.6 describes an **alternating** period-4 cycle when $x_0 = 5$, $a_0 = 0.4$ and $a_1 = -2.5$.

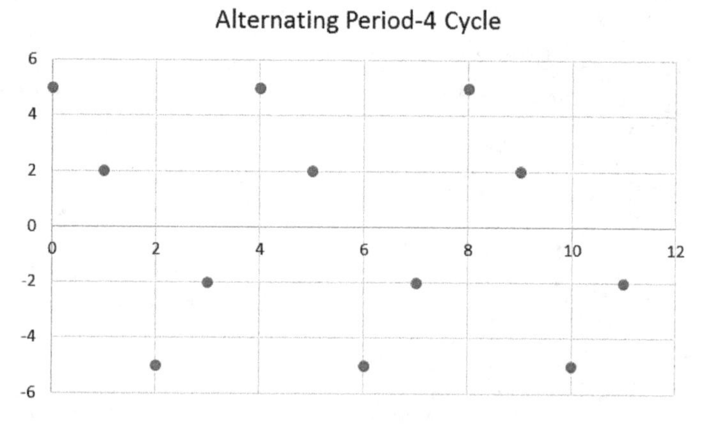

Figure 7.6: An alternating period-4 cycle.

If $P = -1$ then every solution of Eq. (7.6) is the corresponding alternating period-4 cycle:

$$x_0,\ a_0x_0,\ -x_0,\ -a_0x_0,\ \ldots, \tag{7.9}$$

where $x_{n+2} = -x_n$ for all $n \geq 0$. Figures 7.5 and 7.6 trace a **mirror image** of their alternating terms. Figure 7.5 is a mirror image of the lower step and the upper step while Figure 7.6 describes a mirror image of the **descending terms** and the **ascending terms**.

Next, assume that $\{a_n\}_{n=0}^{\infty}$ is a period-3 sequence and formulate the **explicit solution** of Eq. (7.6) and specify the periodicity character.

By iterations and induction you get:

$$\begin{aligned}
&x_0,\\
&x_1 = a_0x_0,\\
&x_2 = a_1x_1 = a_1a_0x_0,\\
&x_3 = a_2x_2 = [a_2a_1a_0]\,x_0,\\
&x_4 = a_0x_3 = a_0\,[a_2a_1a_0]\,x_0,\\
&x_5 = a_1x_4 = a_1a_0\,[a_2a_1a_0]\,x_0,\\
&x_6 = a_2x_5 = [a_2a_1a_0]^2x_0,\\
&\vdots
\end{aligned} \tag{7.10}$$

Let $P = a_0a_1a_2$. Via (7.10), for all $n \in \mathbb{N}$ you formulate the cognate solution to Eq. (7.6):

$$\begin{cases}
x_{3n} = P^nx_0,\\
x_{3n+1} = a_0P^nx_0,\\
x_{3n+2} = a_0a_1P^nx_0.
\end{cases} \tag{7.11}$$

Via (7.11), every solution of Eq. (7.6) is periodic with:

(i) Period-3 if $P = 1$.
(ii) Period-6 if $P = -1$.

Figure 7.7 resembles a "**descending step-shaped**" period-3 cycle when $x_0 = 2$, $a_0 = 1$, $a_1 = 0.25$ and $a_2 = 4$.

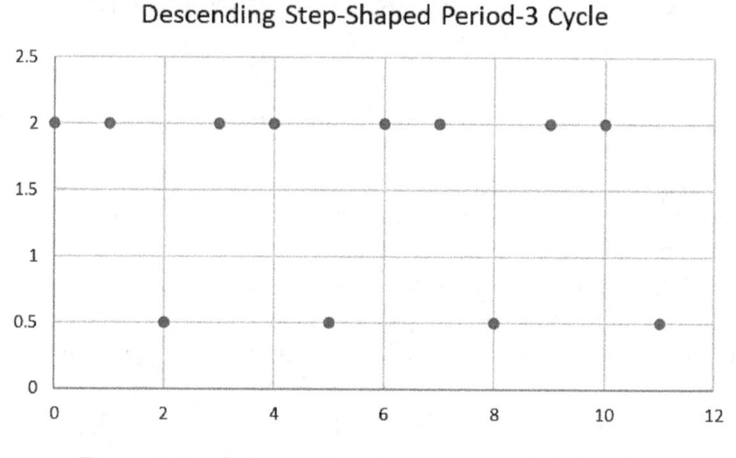

Figure 7.7: A descending step-shaped period-3 cycle.

Similarly, you can select values of x_0, a_0, a_1 and a_2 to generate period-3 cycles with numerous shapes. Figure 7.8 traces an **increasing period-3 cycle** when $x_0 = -3$, $a_0 = 0.8$, $a_1 = -0.5$ and $a_2 = -2.5$.

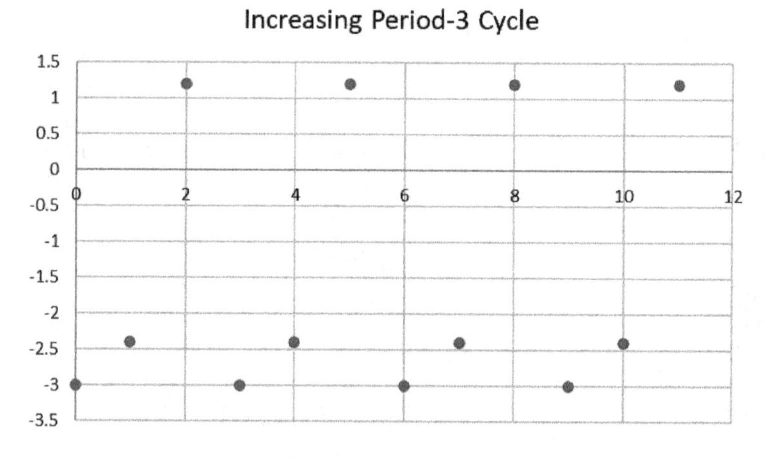

Figure 7.8: An increasing period-3 cycle.

Figure 7.8 depicts an increasing period-3 cycle with two negative terms and a positive term. The next graph renders an **alternating period-6 cycle** when $x_0 = 2$, $a_0 = 0.8$, $a_1 = 0.25$ and $a_2 = -5$.

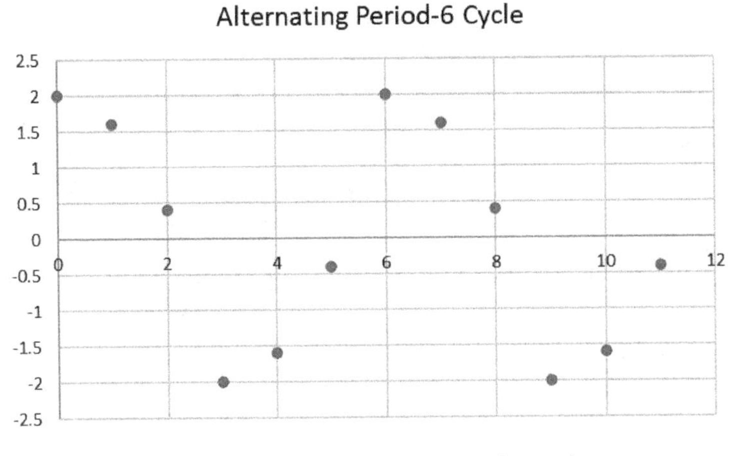

Figure 7.9: An alternating period-6 cycle.

Figure 7.9 depicts an alternating period-6 cycle with three positive descending terms and three negative ascending terms. In addition, if $P = -1$ then every solution to Eq. (7.6) is the corresponding alternating period-6 cycle:

$$x_0,\ a_0x_0,\ a_0a_1x_0,\ -x_0,\ -a_0x_0,\ -a_0a_1x_0,\ \ldots, \tag{7.12}$$

where $x_{n+3} = -x_n$ for all $n \geq 0$.

Next, assume that $\{a_n\}_{n=0}^{\infty}$ is a period-4 sequence and let $P = a_0a_1a_2a_3$. Then parallel to (7.8) and (7.11), for all $n \in \mathbb{N}$ you formulate the cognate solution to Eq. (7.6):

$$\begin{cases} x_{4n} = P^n x_0, \\ x_{4n+1} = a_0P^n x_0, \\ x_{4n+2} = a_0a_1P^n x_0, \\ x_{4n+3} = a_0a_1a_2P^n x_0. \end{cases} \tag{7.13}$$

Via (7.13), every solution of Eq. (7.6) is periodic with:

(i) Period-4 if $P = 1$.
(ii) Period-8 if $P = -1$.

Figure 7.10 describes a **descending period-4** cycle with one positive term and three negative terms when $x_0 = 1$, $a_0 = -1$, $a_1 = 2$, $a_2 = 4$ and $a_3 = -0.125$.

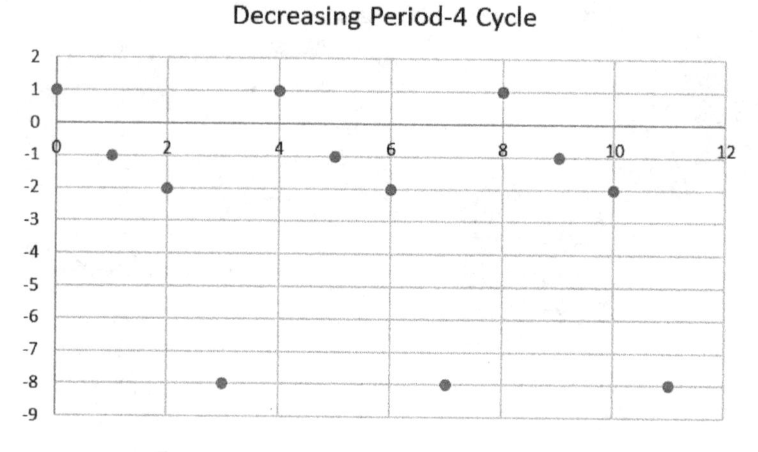

Figure 7.10: A decreasing period-4 cycle.

Figure 7.11 presents an **alternating period-8 cycle** when $x_0 = 1$, $a_0 = 1$, $a_1 = 2$, $a_2 = 2$ and $a_3 = -0.25$.

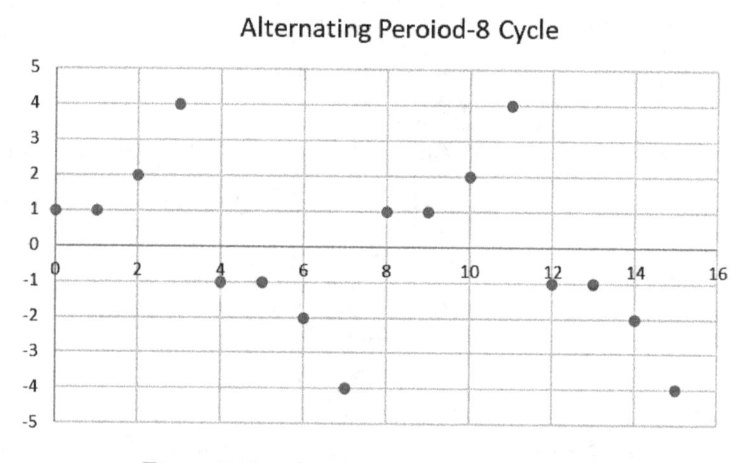

Figure 7.11: An alternating period-8 cycle.

In comparison with Figures 7.6 and 7.9, Figure 7.11 resembles a mirror image of an alternating period-8 cycle with positive increasing terms and their corresponding negative decreasing terms. Analogous to (7.9) and (7.12), if $P = -1$ then every solution of Eq. (7.6) is the corresponding alternating period-8 cycle:

$$x_0,\ a_0x_0,\ a_0a_1x_0,\ a_0a_1a_2x_0,\ -x_0,\ -a_0x_0,\ -a_0a_1x_0,\ -a_0a_1a_2x_0,\ \ldots, \tag{7.14}$$

where $x_{n+4} = -x_n$ for all $n \geq 0$.

Next, suppose that $\{a_n\}_{n=0}^{\infty}$ is a period-k sequence ($k \geq 2$) and let

$$P = \prod_{i=0}^{k-1} a_i.$$

Then analogous to (7.8), (7.11), and (7.13), you formulate the associated solution to Eq. (7.6):

$$\begin{cases} x_{kn} = P^n x_0, \\ x_{kn+1} = a_0 P^n x_0, \\ x_{kn+2} = a_0 a_1 P^n x_0, \\ \vdots \\ x_{kn+k-2} = \left[\prod_{i=0}^{k-3} a_i\right] P^n x_0, \\ x_{kn+k-1} = \left[\prod_{i=0}^{k-2} a_i\right] P^n x_0. \end{cases} \tag{7.15}$$

Hence, via (7.15), every solution of Eq. (7.6) is periodic with:

(i) Period-k if $P = 1$.
(ii) Period-$2k$ if $P = -1$.

You can conclude that when $P = -1$ then every solution of Eq. (7.6) is an **alternating period-2k cycle**, where $x_{n+k} = -x_n$ for all $n \geq 0$. The next section will examine the periodic traits of Eq. (7.1) in the **additive form**.

7.3 Additive Form of Eq. (7.1)

In this section, the goal is to analyze and formulate the periodic traits of Eq. (7.1) in the corresponding **additive form**:

$$x_{n+1} = x_n + b_n, \quad n = 0, 1, \ldots, \tag{7.16}$$

where $\{b_n\}_{n=0}^{\infty}$ is a period-k sequence ($k \geq 2$). First assume that $\{b_n\}_{n=0}^{\infty}$ as a period-2 sequence and formulate the **explicit solution** of Eq. (7.16) and specify the periodicity character.

By induction and iterations you acquire:

$$\begin{aligned}
&x_0,\\
&x_1 = x_0 + b_0,\\
&x_2 = x_1 + b_1 = x_0 + [b_0 + b_1],\\
&x_3 = x_2 + b_0 = x_0 + b_0 + [b_0 + b_1],\\
&x_4 = x_3 + b_1 = x_0 + 2\,[b_0 + b_1],\\
&x_5 = x_4 + b_0 = x_0 + b_0 + 2\,[b_0 + b_1],\\
&x_6 = x_5 + b_1 = x_0 + 3\,[b_0 + b_1],\\
&\vdots
\end{aligned} \tag{7.17}$$

Let $S = b_0 + b_1$. Via (7.17), for all $n \in \mathbb{N}$ you obtain:

$$\begin{cases} x_{2n} = x_0 + Sn, \\ x_{2n+1} = x_0 + b_0 + Sn. \end{cases} \tag{7.18}$$

Hence, via (7.17) and (7.18), you see that every solution of Eq. (7.16) is periodic with period-2 provided that $S = 0$. Figure 7.12 sketches a **decreasing period-2 cycle** when $x_0 = 4$, $b_0 = -5$ and $b_1 = 5$.

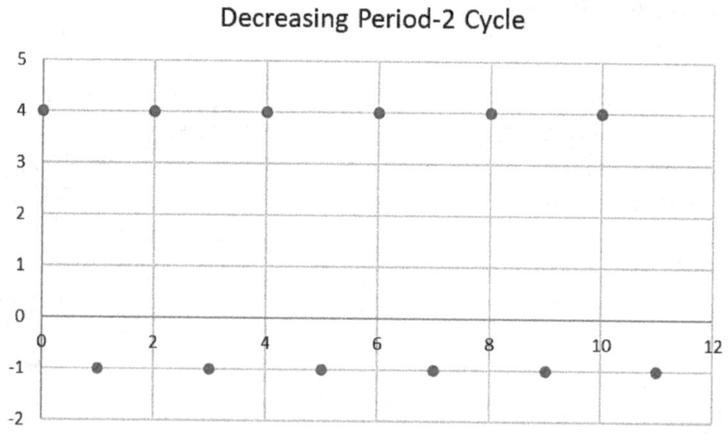

Decreasing Period-2 Cycle

Figure 7.12: A decreasing period-2 cycle.

Figure 7.13 traces an **increasing positive period-2 cycle** when $x_0 = 1$, $b_0 = 4$ and $b_1 = -4$.

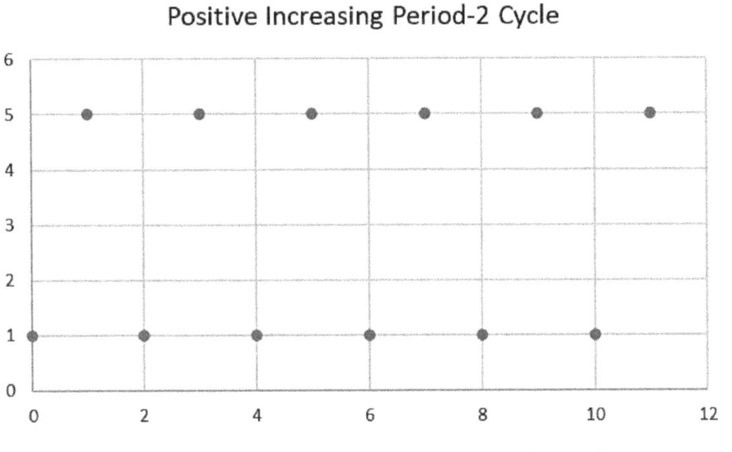

Figure 7.13: A positive increasing period-2 cycle.

Via (7.18), every solution of Eq. (7.6) is periodic with period-2 if $S = 0$ with the corresponding pattern:

$$x_0,\ x_0 + b_0,\ x_0,\ x_0 + b_0,\ \ldots. \tag{7.19}$$

By selecting different specific values of x_0, b_0 and b_1, you can obtain either a positive periodic cycle as in Figure 7.13, a negative periodic cycle or a periodic cycle with a positive and a negative term as in Figure 7.12.

Next, assume that $\{b_n\}_{n=0}^{\infty}$ is a period-3 sequence and formulate the **explicit solution** of Eq. (7.16) and specify the periodicity character.

Now let:

$$S = b_0 + b_1 + b_2.$$

Then parallel to (7.17), for all $n \in \mathbb{N}$ you formulate the cognate solution of Eq. (7.16):

$$\begin{cases} x_{3n} = x_0 + Sn, \\ x_{3n+1} = x_0 + b_0 + Sn, \\ x_{3n+2} = x_0 + b_0 + b_1 + Sn. \end{cases} \tag{7.20}$$

Via (7.20), every solution of Eq. (7.16) is periodic with period-3 if $S = 0$ with the corresponding pattern:

$$x_0,\ x_0 + b_0,\ x_0 + b_0 + b_1,\ x_0,\ x_0 + b_0,\ x_0 + b_0 + b_1,\ \ldots. \tag{7.21}$$

Figure 7.14 presents a **positive decreasing period-3 cycle** when $x_0 = 5$, $b_0 = -2$, $b_1 = -2$ and $b_2 = 4$.

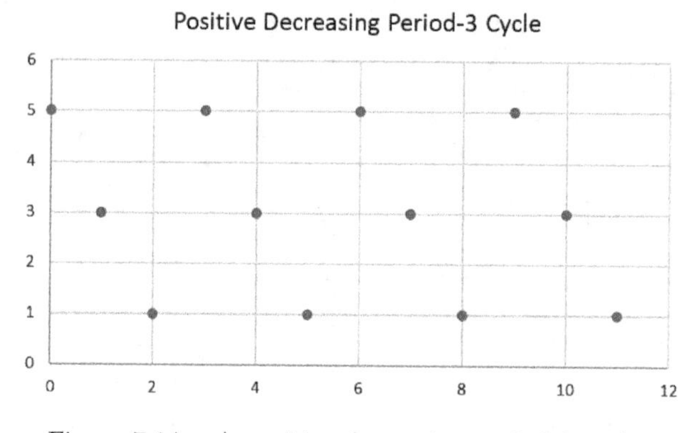

Figure 7.14: A positive decreasing period-3 cycle.

Switching the negative sign of either b_0, b_1 or b_2 will affect the shape of the periodic cycle. Figure 7.15 portrays a **positive increasing period-3 cycle** with $x_0 = 1$, $b_0 = 1$, $b_1 = 3$ and $b_2 = -4$.

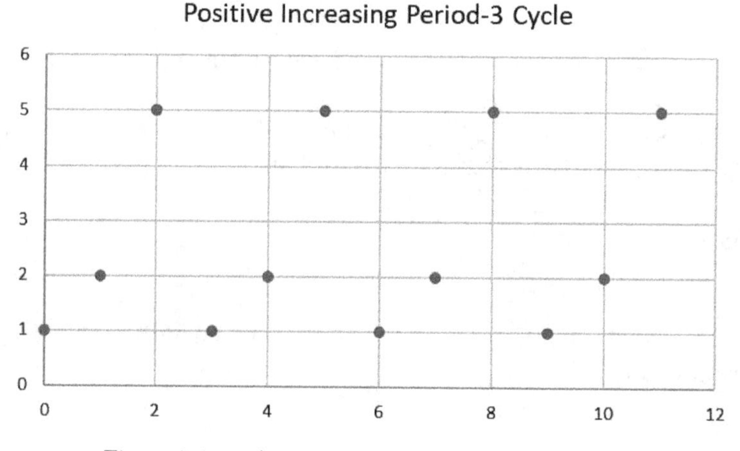

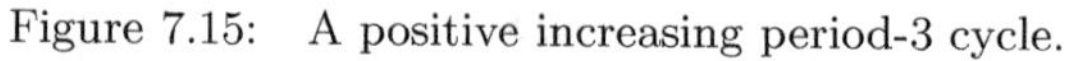

Figure 7.15: A positive increasing period-3 cycle.

Next, assume that $\{b_n\}_{n=0}^{\infty}$ is a period-4 sequence and let:

$$S = b_0 + b_1 + b_2 + b_3.$$

Then parallel to (7.18) and (7.20), for all $n \in \mathbb{N}$ you formulate the cognate solution to Eq. (7.16):

$$\begin{cases} x_{4n} = x_0 + Sn, \\ x_{4n+1} = x_0 + b_0 + Sn, \\ x_{4n+2} = x_0 + b_0 + b_1 + Sn, \\ x_{4n+3} = x_0 + b_0 + b_1 + b_2 + Sn. \end{cases} \tag{7.22}$$

Via (7.22), every solution of Eq. (7.16) is periodic with period-4 if $S = 0$ with the corresponding pattern:

$$x_0, \ x_0 + b_0, \ x_0 + b_0 + b_1, \ x_0 + b_0 + b_1 + b_2, \ \ldots. \tag{7.23}$$

Figure 7.16 traces a **positive descending period-4 cycle** when $x_0 = 8$, $b_0 = -1$, $b_1 = -2$, $b_2 = -3$ and $b_3 = 6$.

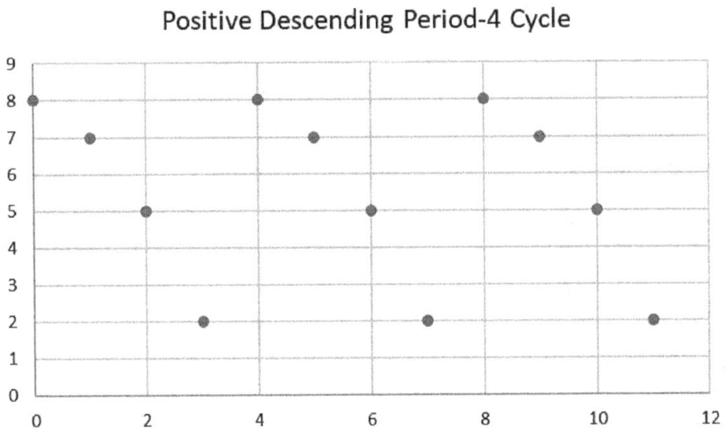

Positive Descending Period-4 Cycle

Figure 7.16: A positive descending period-4 cycle.

On the other hand, Figure 7.17 presents an **ascending period-4 cycle** with two negative terms and two positive terms when $x_0 = 2$, $b_0 = 1$, $b_1 = 2$, $b_2 = 3$ and $b_3 = -6$.

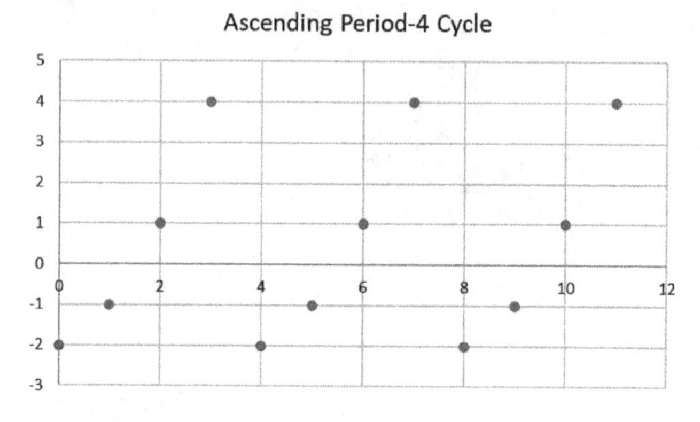

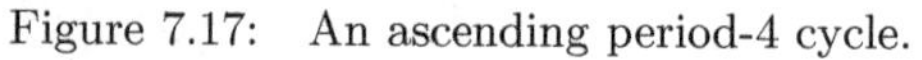
Figure 7.17: An ascending period-4 cycle.

Next, suppose that $\{b_n\}_{n=0}^{\infty}$ is a period-k sequence ($k \geq 2$) and let:

$$S = \sum_{i=0}^{k-1} b_i.$$

Analogous to (7.18), (7.20), and (7.22), for all $n \in \mathbb{N}$ you formulate the cognate solution to Eq. (7.16):

$$\begin{cases} x_{kn} = x_0 + Sn, \\ x_{kn+1} = x_0 + b_0 + Sn, \\ x_{kn+2} = x_0 + b_0 + b_1 + Sn, \\ \vdots \\ x_{kn+k-2} = x_0 + \left[\sum_{i=0}^{k-3} b_i\right] + Sn, \\ x_{kn+k-1} = x_0 + \left[\sum_{i=0}^{k-2} b_i\right] + Sn. \end{cases} \tag{7.24}$$

Via (7.24), every solution of Eq. (7.16) is periodic with period-k if $S = 0$.

7.3.1 *Special Case of Additive Form of Eq. (7.1)*

This section's aims are to examine and formulate the periodic traits of Eq. (7.1) in the corresponding **additive form**:

$$x_{n+1} = -x_n + b_n, \quad n = 0, 1, \ldots, \tag{7.25}$$

where $\{b_n\}_{n=0}^{\infty}$ is a period-k sequence ($k \geq 2$) and compare the similarities and differences with the periodic features of Eq. (7.16).

First, your aim is to examine the periodicity character of Eq. (7.25) when $\{b_n\}_{n=0}^{\infty}$ as a period-2 sequence. You will show that Eq. (7.25) has no period-2 cycles.

Suppose that $x_2 = x_0$. Then by iterations you get:

$$\begin{aligned} &x_0, \\ &x_1 = -x_0 + b_0, \\ &x_2 = -x_1 + b_1 = -[x_0 + b_0] + b_1 = x_0 - b_0 + b_1 = x_0. \end{aligned} \tag{7.26}$$

Thus, via (7.26) $x_2 = x_0$ if and only if $b_0 = b_1$. This is a contradiction as you assumed that $\{b_n\}_{n=0}^{\infty}$ is a period-2 sequence where $b_0 \neq b_1$.

The next two Sections 7.3.2 and 7.3.3, will signify the contrasting periodic traits of Eq. (7.25) when $\{b_n\}_{n=0}^{\infty}$ is an even-ordered periodic sequence and an odd-ordered periodic sequence. The upcoming section will examine the periodic attributes of Eq. (7.25) when $\{b_n\}_{n=0}^{\infty}$ is an odd-ordered periodic sequence.

7.3.2 $\{b_n\}_{n=0}^{\infty}$ *is an Odd-Ordered Periodic Sequence*

This section will focus on unique periodic features of Eq. (7.25) when $\{b_n\}_{n=0}^{\infty}$ is an odd-ordered periodic sequence. First, suppose that $\{b_n\}_{n=0}^{\infty}$ is a period-3 sequence and we formulate the **explicit solution** of Eq. (7.25) and specify the periodicity character.

Set $x_3 = x_0$ and by iterations you get:

$$\begin{aligned} &x_0, \\ &x_1 = -x_0 + b_0, \\ &x_2 = -[x_1] + b_1 = -[-x_0 + b_0] + b_1 = x_0 + b_1 - b_0, \\ &x_3 = -[x_2] + b_2 = -[x_0 + b_1 - b_0] + b_2 = -x_0 + b_2 - b_1 + b_0 = x_0. \end{aligned} \tag{7.27}$$

Then via (7.27) you obtain the corresponding initial condition:

$$x_0 = \frac{b_0 - b_1 + b_2}{2}, \tag{7.28}$$

and the cognate **unique** period-3 cycle:

$$\frac{b_0 - b_1 + b_2}{2}, \ \frac{b_1 - b_2 + b_0}{2}, \ \frac{b_2 - b_0 + b_1}{2}, \ \ldots. \tag{7.29}$$

This is the also first time that you encounter a **unique periodic cycle**. Figure 7.18 portrays a "**descending triangular-shaped**" period-3 cycle when $x_0 = 2$, $b_0 = 1$, $b_1 = 1$ and $b_2 = 4$.

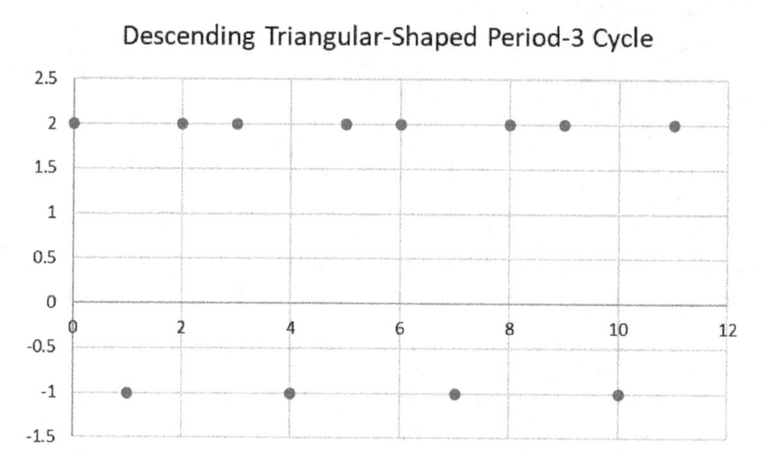

Figure 7.18: A descending triangular-shaped period-3 cycle.

On contrary, Figure 7.19 traces a **descending step-shaped** period-3 cycle when $x_0 = 3$, $b_0 = 6$, $b_1 = 1$ and $b_2 = 1$.

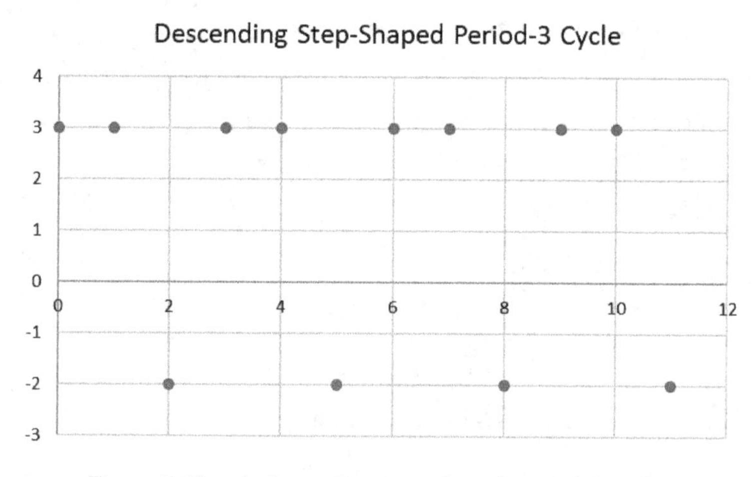

Figure 7.19: A descending step-shaped period-3 cycle.

It is interesting to note that if:

$$x_0 \neq \frac{b_0 - b_1 + b_2}{2}, \tag{7.30}$$

then every solution of Eq. (7.25) is periodic with period-6. This will be left as an end of chapter exercise to verify by showing that $x_6 = x_0$. It is also of paramount interest to examine for what values of x_0, b_0, b_1 and b_2 you obtain an oscillatory period-6 cycle of Eq. (7.25).

Period-6 cycles emerge as 2 and 3 are relatively prime, where $2 \cdot 3 = 6$ and as $\{b_n\}_{n=0}^{\infty}$ is a period-3 sequence. Figure 7.20 resembles an **oscillatory** period-6 cycle when $x_0 = 4$, $b_0 = 1$, $b_1 = 2$ and $b_2 = 3$.

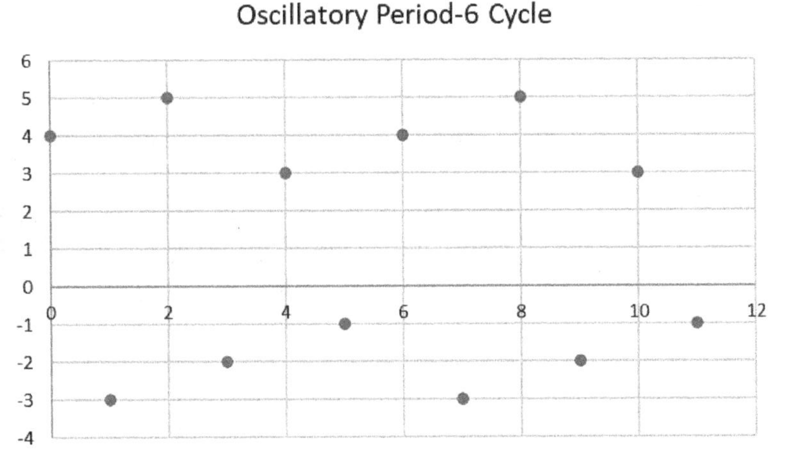

Figure 7.20: An oscillatory period-6 cycle.

Next, assume that $\{b_n\}_{n=0}^{\infty}$ is a period-5 sequence and formulate the **explicit solution** of Eq. (7.25) and specify the periodicity character.

Analogous to (7.27), set $x_5 = x_0$ and by iterations you obtain the corresponding initial condition:

$$x_0 = \frac{b_0 - b_1 + b_2 - b_3 + b_4}{2}. \tag{7.31}$$

In the sigma notation, reformulate (7.31) as the corresponding alternating summation:

$$x_0 = \frac{\sum_{i=0}^{4}(-1)^i\, b_i}{2}. \tag{7.32}$$

You then acquire the cognate **unique** period-5 cycle:

$$\begin{aligned} x_0 &= \frac{b_0 - b_1 + b_2 - b_3 + b_4}{2}, \\ x_1 &= \frac{b_1 - b_2 + b_3 - b_4 + b_0}{2}, \\ x_2 &= \frac{b_2 - b_3 + b_4 - b_0 + b_1}{2}, \\ x_3 &= \frac{b_3 - b_4 + b_0 - b_1 + b_2}{2}, \\ x_4 &= \frac{b_4 - b_0 + b_1 - b_2 + b_3}{2}. \end{aligned} \tag{7.33}$$

For all $j \in [0, 1, 2, 3, 4]$, you can reformulate (7.33) as:

$$x_j = \frac{b_j - b_{j+1} + b_{j+2} - b_{j+3} + b_{j+4}}{2}, \tag{7.34}$$

Figure 7.21 describes a **descending step-shaped** period-5 cycle when $x_0 = 4$, $b_0 = 8$, $b_1 = 6$, $b_2 = 4$, $b_2 = 2$ and $b_3 = 4$.

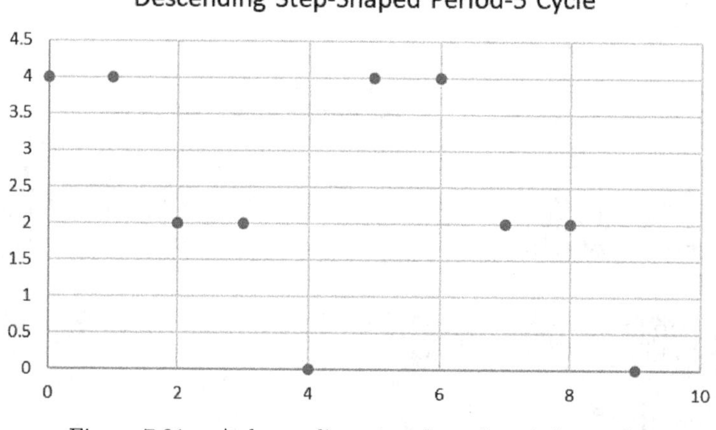

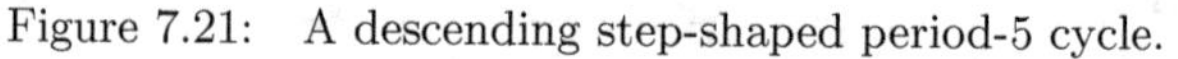
Figure 7.21: A descending step-shaped period-5 cycle.

Analogous to (7.30), if:

$$x_0 \neq \frac{\sum_{i=0}^{4}(-1)^i \, b_i}{2}. \tag{7.35}$$

then every solution of Eq. (7.25) is periodic with period-10. This will be left as an end of chapter exercise to verify by showing that $x_{10} = x_0$. It is also of paramount interest to examine for what values of x_0, b_0, b_1, b_2, b_3 and b_4 you obtain an oscillatory period-10 cycle of Eq. (7.25).

Period-10 cycles emerge as 2 and 5 are relatively prime, where $2 \cdot 5 = 10$ and as $\{b_n\}_{n=0}^{\infty}$ is a period-5 sequence. Figure 7.22 presents an **oscillatory** period-10 cycle when $x_0 = 4$, $b_0 = 1$, $b_1 = 2$, $b_2 = 3$, $b_3 = 2$, and $b_4 = 1$.

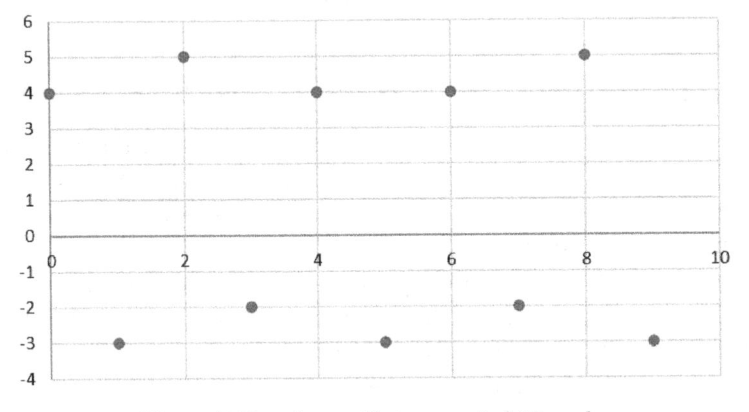

Figure 7.22: An oscillatory period-10 cycle.

Equations (7.28), (7.29), (7.32), and (7.34) then guide you to the examination and the formulation of the unique periodic character of Eq. (7.25) when $\{b_n\}_{n=0}^{\infty}$ is an odd-ordered periodic sequence. In fact, suppose that $\{b_n\}_{n=0}^{\infty}$ is a period-$(2k+1)$ sequence, $(k \in \mathbb{N})$. Then Eq. (7.25) has a **unique periodic cycle** with period-$(2k+1)$ where:

$$x_0 = \frac{\sum_{i=0}^{2k}(-1)^i \, b_i}{2}, \tag{7.36}$$

and for all $j \in [0, 1, 2, \ldots, 2k]$, the cognate period-$(2k+1)$ cycle:

$$x_j = \frac{\sum_{i=0}^{2k}(-1)^i \, b_{i+j}}{2}, \tag{7.37}$$

First, (7.36) transitions from (7.28) and (7.32). Via (7.28) and (7.31) you can conclude that (7.36) has $k+1$ terms with a positive sign and k terms with a negative sign. In fact, the even-indexed coefficients have a positive sign while the odd-indexed coefficients have a negative sign. Analogous to (7.30) and (7.35), the next interesting periodic feature of Eq. (7.25) is that every solution of Eq. (7.25) is periodic with period with period-$2(2k+1)$ if:

$$x_0 \neq \frac{\sum_{i=0}^{2k}(-1)^i \, b_i}{2}, \tag{7.38}$$

and will be left as end of chapter exercise to verify. The upcoming section will examine the periodic attributes of Eq. (7.25) when $\{b_n\}_{n=0}^{\infty}$ an even-ordered periodic sequence.

7.3.3 $\{b_n\}_{n=0}^{\infty}$ *is an Even-Ordered Periodic Sequence*

This section will focus on the examination and formulation of periodic attributes of Eq. (7.25) when $\{b_n\}_{n=0}^{\infty}$ is an even-ordered period-$2k$ sequence, $(k \geq 2)$. You will discover the necessary and sufficient criteria for existence of period-$2k$ cycles with various shapes.

The period-$2k$ cycles Eq. (7.25) and their shapes will not be unique and will depend on the initial condition x_0 and on the relationship between the parameters of the sequence $\{b_n\}_{n=0}^{\infty}$. You will first examine the periodic features of Eq. (7.25) when $\{b_n\}_{n=0}^{\infty}$ is a period-4 sequence and formulate the **explicit solution** of Eq. (7.25) and specify the periodicity character.

By iteration you procure:

$$\begin{aligned}
&x_0,\\
&x_1 = -x_0 + b_0,\\
&x_2 = -[x_1] + b_1 = -[-x_0 + b_0] + b_1 = x_0 + b_1 - b_0,\\
&x_3 = -[x_2] + b_2 = -[x_0 + b_1 - b_0] + b_2 = -x_0 + b_2 + b_0 - b_1\\
&x_4 = -[x_3] + b_3 = -[-x_0 + b_2 + b_0 - b_1] + b_3,\\
&\qquad\quad = x_0 - [b_2 + b_0] + [b_1 + b_3] = x_0.
\end{aligned} \tag{7.39}$$

Via (7.39), period-4 cycles exist if and only if:

$$b_1 + b_3 = b_0 + b_2, \tag{7.40}$$

with the corresponding period-4 pattern:

$$x_0, \ -x_0 + b_0, \ x_0 + b_1 - b_0, \ -x_0 + b_2 + b_0 - b_1, \ \ldots. \tag{7.41}$$

Note that the period-4 cycle described by (7.41) is not unique and depends on the initial condition x_0. Graphically, (7.41) can emerge in various shapes such as ascending shaped, descending shaped, ascending step-shaped, descending step-shaped, etc. The cycles' shapes will not only depend on the initial condition x_0 but also depend on the relationship between the parameters b_0, b_1, b_2 and b_3.

Figure 7.23 sketches an **ascending step-shaped** period-4 cycle with two repeated negative terms and two positive terms when $x_0 = -3$, $b_0 = -6$, $b_1 = -2$, $b_2 = 6$ and $b_3 = 2$.

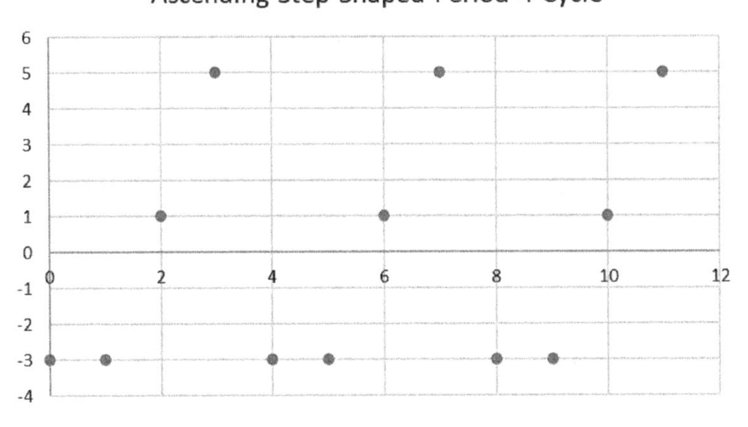

Figure 7.23: An ascending step-shaped period-4 cycle.

On the contrary, Figure 7.24 resembles a **descending step-shaped** period-4 cycle with three positive term and a 0 term when $x_0 = 4$, $b_0 = 6$, $b_1 = 4$, $b_2 = 2$ and $b_3 = 4$.

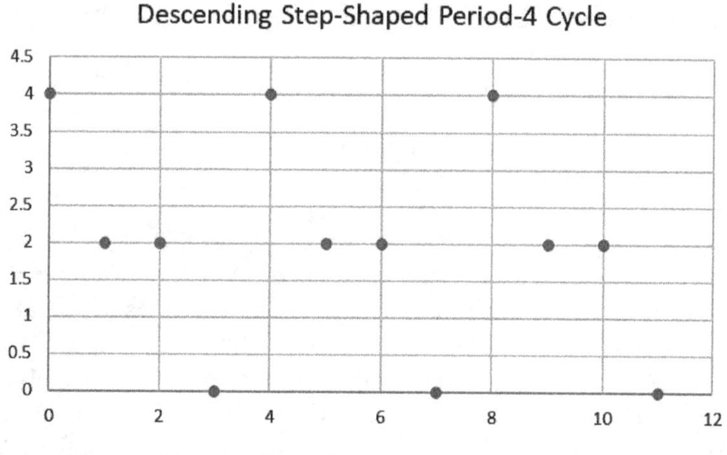

Figure 7.24: An descending step-shaped period-4 cycle.

Figure 7.25 renders a **trapezoidal-shaped** period-4 cycle with two negative terms and two positive terms when $x_0 = -1$, $b_0 = 3$, $b_1 = 8$, $b_2 = 3$ and $b_3 = -2$.

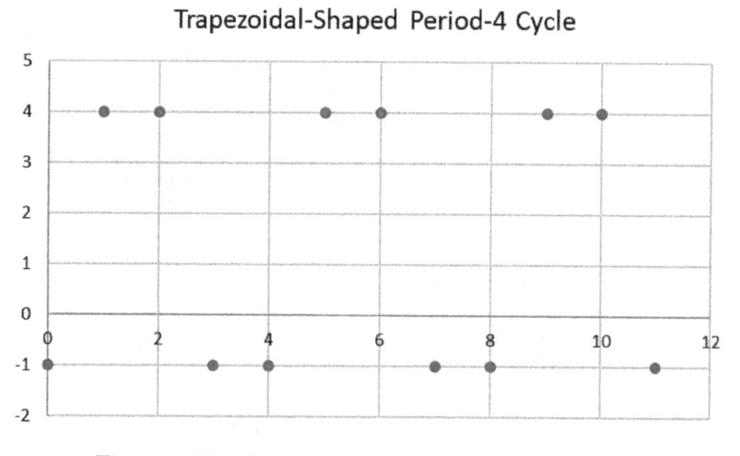

Figure 7.25: A trapezoidal-shaped period-4 cycle.

This then guides you to the upcoming result that portrays the periodic characteristics of Eq. (7.25) when $\{b_n\}_{n=0}^{\infty}$ is an even-ordered periodic sequence. Suppose that $\{b_n\}_{n=0}^{\infty}$ is a period-$2k$ sequence, $(k \geq 2)$. Then every solution of Eq. (7.25) is periodic with period-$2k$

if and only if:

$$\sum_{i=1}^{k} b_{2i-1} = \sum_{i=1}^{k} b_{2i-2}. \tag{7.42}$$

Note that (7.42) is a direct consequence of (7.40), where the sum of all the odd-indexed coefficients must be equal to the sum of all the even-indexed coefficients. Furthermore, (7.42) has k terms on the left side and k terms on the right side. The proof of (7.42) will be left as an end of chapter exercise.

The next section of this chapter will examine and formulate the periodic characteristics of Eq. (7.1) in the **additive and multiplicative forms** where $\{a_n\}_{n=0}^{\infty}$ and $\{b_n\}_{n=0}^{\infty}$ are periodic sequences of real numbers.

7.4 Additive and Multiplicative Forms of Eq. (7.1)

In this section, you will analyze and formulate the periodic features of Eq. (7.1) in the **additive and multiplicative forms** where $\{a_n\}_{n=0}^{\infty}$ and $\{b_n\}_{n=0}^{\infty}$ are periodic sequences of real numbers. You will examine two cases when $\{a_n\}_{n=0}^{\infty}$ and $\{b_n\}_{n=0}^{\infty}$ are periodic with the same period or periodic with different periods. The consequent section will assume that $\{a_n\}_{n=0}^{\infty}$ and $\{b_n\}_{n=0}^{\infty}$ are periodic with the same period.

7.4.1 $\{a_n\}_{n=0}^{\infty}$ *and* $\{b_n\}_{n=0}^{\infty}$ *are the Same Period*

In this section, assume that $\{a_n\}_{n=0}^{\infty}$ and $\{b_n\}_{n=0}^{\infty}$ are periodic with the same period-k, ($k \geq 2$) and you will encounter unique periodic cycles of Eq. (7.1). The first aim is to outline the periodic features of Eq. (7.1) when $\{a_n\}_{n=0}^{\infty}$ and $\{b_n\}_{n=0}^{\infty}$ are period-2 sequences and formulate the **explicit solution** of Eq. (7.1) and specify the periodicity character.

First, set $x_2 = x_0$ and by iterations you get:

$$x_0,$$
$$x_1 = a_0 x_0 + b_0,$$
$$x_2 = a_1 [x_1] + b_1 = a_1 [a_0 x_0 + b_0] + b_1 = a_0 a_1 x_0 + a_1 b_0 + b_1 = x_0. \tag{7.43}$$

Via (7.43), you obtain the corresponding initial condition:

$$x_0 = \frac{a_1 b_0 + b_1}{1 - a_0 a_1}, \tag{7.44}$$

provided that $a_0 a_1 \neq 1$ and the associated **unique period-2 cycle**:

$$\frac{a_1 b_0 + b_1}{1 - a_0 a_1}, \frac{a_0 b_1 + b_0}{1 - a_0 a_1}, \frac{a_1 b_0 + b_1}{1 - a_0 a_1}, \frac{a_0 b_1 + b_0}{1 - a_0 a_1}, \ldots. \tag{7.45}$$

First, you acquire a **unique period-2 cycle**. The indices of a_0, a_1, b_0, and b_1 in the numerator shift by an index of one from neighbor to neighbor, while the terms in the denominator do not change. The next two diagrams render various shapes of period-2 cycles that vary on the relationship between the terms of the sequences $\{a_n\}_{n=0}^{\infty}$ and $\{b_n\}_{n=0}^{\infty}$ and the initial condition x_0.

Figure 7.26 presents (7.45) as a **descending period-2 cycle** with a positive term and a negative term when $x_0 = 3$, $a_0 = -1$, $a_1 = 1$, $b_0 = 2$ and $b_0 = 4$.

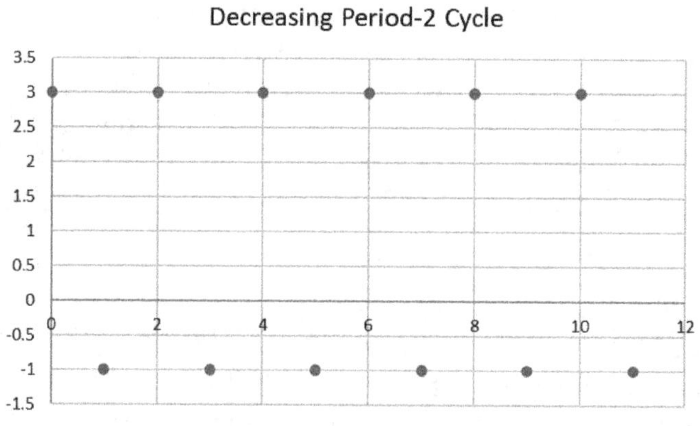

Decreasing Period-2 Cycle

Figure 7.26: A descending period-2 cycle.

On the contrary, Figure 7.27 describes (7.45) as a **positive ascending period-2 cycle** when $x_0 = 1.5$, $a_0 = 1$, $a_1 = -1$, $b_0 = 1$ and $b_0 = 4$.

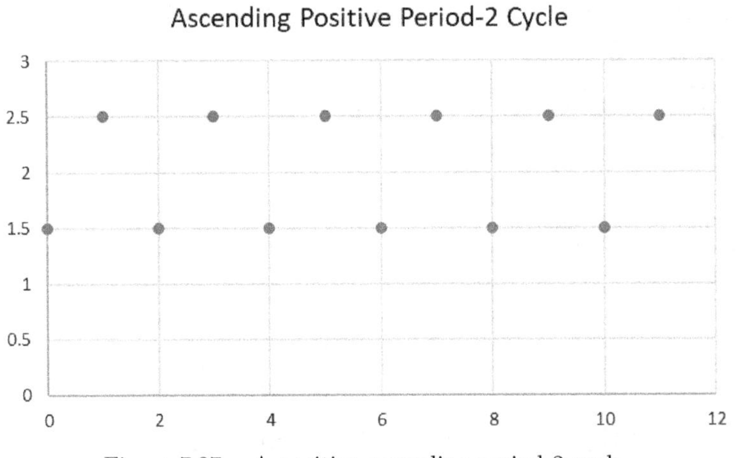

Figure 7.27: A positive ascending period-2 cycle.

Next, assume that $\{a_n\}_{n=0}^{\infty}$ and $\{b_n\}_{n=0}^{\infty}$ are period-3 sequences and formulate the **explicit solution** of Eq. (7.1) and specify the periodicity character.

First, set $x_3 = x_0$ and by iterations you get:

$$\begin{aligned}
&x_0,\\
&x_1 = a_0x_0 + b_0,\\
&x_2 = a_1\left[x_1\right] + b_1 = a_1\left[a_0x_0 + b_0\right] + b_1 = a_0a_1x_0 + a_1b_0 + b_1,\\
&x_3 = a_2\left[x_2\right] + b_2 = a_2\left[a_0a_1x_0 + a_1b_0 + b_1\right] + b_2,\\
&\qquad\qquad\qquad = a_0a_1a_2x_0 + a_1a_2b_0 + a_2b_1 + b_2 = x_0.
\end{aligned} \tag{7.46}$$

Via (7.46), you obtain the corresponding initial condition:

$$x_0 = \frac{a_1a_2b_0 + a_2b_1 + b_2}{1 - a_0a_1a_2}, \tag{7.47}$$

provided that $a_0a_1a_2 \neq 1$. In the **factored form** and **nested form**, we reformulate (7.47) as:

$$x_0 = \frac{a_2\left[a_1b_0 + b_1\right] + b_2}{1 - a_0a_1a_2}, \tag{7.48}$$

and then acquire the associated **unique period-3 cycle:**

$$\frac{a_2\left[a_1b_0+b_1\right]+b_2}{1-a_0a_1a_2},\ \frac{a_0\left[a_2b_1+b_2\right]+b_0}{1-a_0a_1a_2},\ \frac{a_1\left[a_0b_2+b_0\right]+b_1}{1-a_0a_1a_2},\ \ldots. \tag{7.49}$$

For all $j \in [0, 1, 2]$, you can reformulate (7.49) as:

$$x_j = \frac{a_{2+j}\left[a_{1+j}b_j + b_{1+j}\right] + b_{2+j}}{1 - a_0a_1a_2}. \tag{7.50}$$

The terms of Equation (7.49) or (7.50) can emerge in various shapes depending on x_0, a_0, a_1, a_2, b_0, b_1 and b_2. Figure 7.28 resembles a **descending step-shaped** period-3 cycle with one a positive term and two negative equal terms when $x_0 = 2$, $a_0 = -1$, $a_1 = 2$, $a_2 = 2$, $b_0 = 1$, $b_1 = 1$ and $b_2 = 4$.

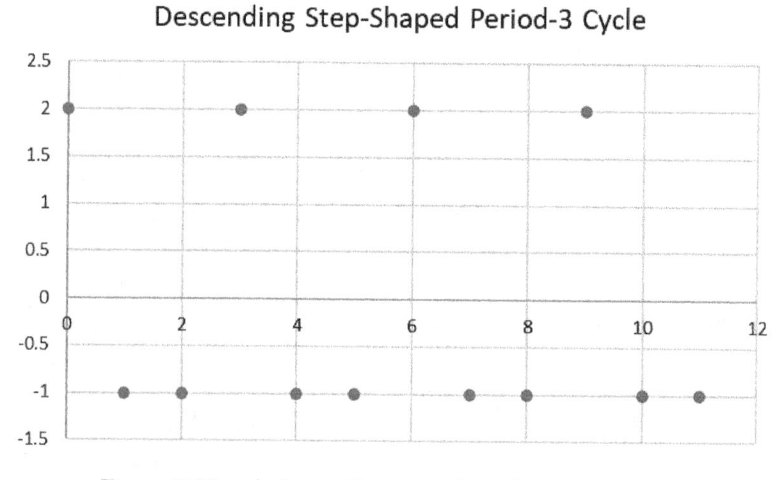

Figure 7.28: A descending step-shaped period-3 cycle.

Figure 7.29 portrays a **non-negative descending step-shaped** period-3 cycle with two positive equal terms and a 0 term when $x_0 = 2$, $a_0 = 0.5$, $a_1 = -1$, $a_2 = 2$, $b_0 = 1$, $b_1 = 2$ and $b_2 = 2$.

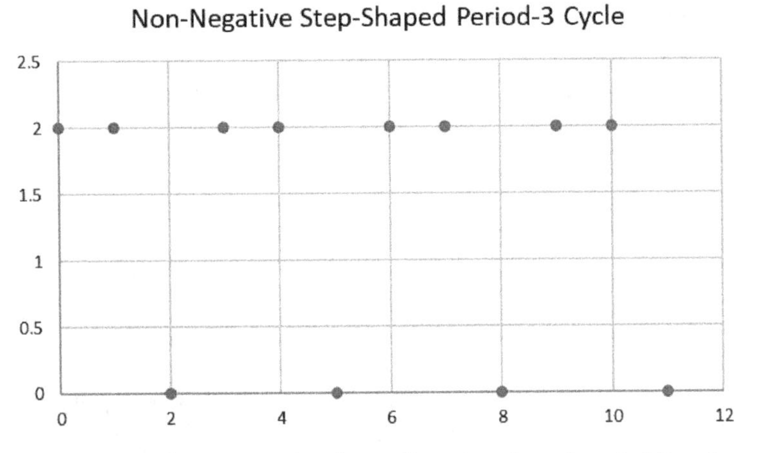

Figure 7.29: A non-negative descending step-shaped period-3 cycle.

Figure 7.30 presents a **negative ascending** period-3 cycle with three negative terms when $x_0 = -6$, $a_0 = 1$, $a_1 = 1$, $a_2 = 2$, $b_0 = 1$, $b_1 = 1$ and $b_2 = 2$.

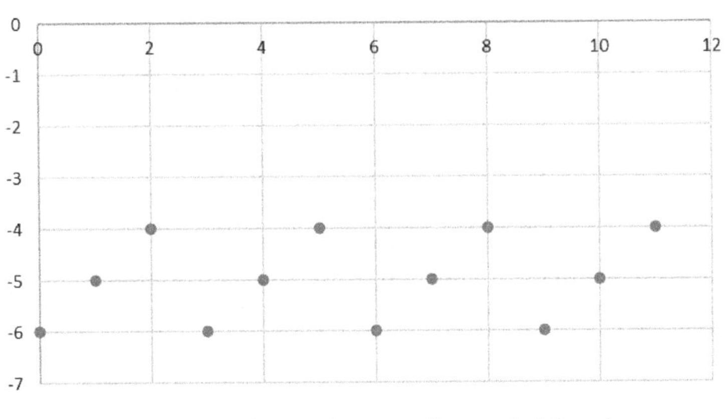

Figure 7.30: A negative ascending period-3 cycle.

Equations (7.44), (7.47) and (7.61) then guide you to the examination and the formulation of the unique periodic character of Eq. (7.1) when $\{a_n\}_{n=0}^{\infty}$ and $\{b_n\}_{n=0}^{\infty}$ are period-k sequences ($k \geq 2$). In fact, suppose that $\{a_n\}_{n=0}^{\infty}$ and $\{b_n\}_{n=0}^{\infty}$ are period-k sequences ($k \geq 2$). Then, Eq. (7.1) has a **unique period-k cycle** ($k \geq 2$) with the corresponding initial condition:

$$x_0 = \frac{a_{k-1}\left[\cdots \left[a_3\left[a_2\left[a_1 b_0 + b_1\right] + b_2\right] + b_3\right] \cdots\right] + b_{k-1}}{1 - \left[\prod_{i=0}^{k-1} a_i\right]}, \qquad (7.51)$$

where

$$\prod_{i=0}^{k-1} a_i \neq 1.$$

Equation (7.51) is expressed as a **nested sequence** and you can reformulate (7.51) in the **sigma and product form** as:

$$x_0 = \frac{\sum_{i=0}^{k-2}\left[\prod_{j=i+1}^{k-1} a_j b_i\right] + b_{k-1}}{1 - \left[\prod_{i=0}^{k-1} a_i\right]}, \qquad (7.52)$$

and for all $j \in [0, 1, 2, \ldots, k-1]$, the cognate period-$k$ cycle:

$$x_j = \frac{a_{k-1+j}\left[\cdots \left[a_{3+j}\left[a_{2+j}\left[a_{1+j} b_{0+j} + b_{1+j}\right] + b_{2+j}\right] + b_{3+j}\right] \cdots\right] + b_{k-1+j}}{1 - \left[\prod_{i=0}^{k-1} a_i\right]}. \qquad (7.53)$$

First, note that (7.51) and (7.52) extend from (7.45), (7.47) and (7.61). Also observe that (7.53) extends from (7.45), (7.49) and (7.50). The next interesting periodic feature of Eq. (7.1) to examine is when

$$x_0 \neq \frac{a_{k-1}\left[\cdots \left[a_3\left[a_2\left[a_1 b_0 + b_1\right] + b_2\right] + b_3\right] \cdots\right] + b_{k-1}}{1 - \left[\prod_{i=0}^{k-1} a_i\right]}, \qquad (7.54)$$

and will be left as end of chapter exercise to verify. The succeeding section will investigate the periodic character of Eq. (7.1) when the periodic sequences $\{a_n\}_{n=0}^{\infty}$ and $\{b_n\}_{n=0}^{\infty}$ are composed of different periods.

7.4.2 $\{a_n\}_{n=0}^{\infty}$ and $\{b_n\}_{n=0}^{\infty}$ are Different Periods

In this section, assume that $\{a_n\}_{n=0}^{\infty}$ and $\{b_n\}_{n=0}^{\infty}$ are periodic with different periods. First, you will address the periodic attributes of Eq. (7.1) when $\{a_n\}_{n=0}^{\infty}$ is periodic with period-2 and $\{b_n\}_{n=0}^{\infty}$ is periodic with period-4 and will show the existence of a unique period-4 cycle.

Set $x_4 = x_0$. Then by iterations you obtain the corresponding initial condition:

$$x_0 = \frac{a_0 a_1^2 b_0 + a_1 b_1 + a_1 b_2 + b_3}{1 - [a_0 a_1]^2}, \tag{7.55}$$

provided that $a_0 a_1 \neq \pm 1$. First, reformulate (7.55) in the **sigma and product form** as:

$$x_0 = \frac{\sum_{i=0}^{2} \left[\prod_{j=i+1}^{3} a_j b_i\right] + b_3}{1 - [a_0 a_1]^2}. \tag{7.56}$$

In the **factored and nested form**, reformulate (7.55) as:

$$x_0 = \frac{a_1 \left[a_0 \left[a_1 b_0 + b_1\right] + b_2\right] + b_3}{1 - [a_0 a_1]^2}, \tag{7.57}$$

Then for all $j \in [0, 1, 2, 3]$, you can formulate the corresponding period-4 cycle of Eq. (7.1):

$$x_j = \frac{a_{1+j} \left[a_j \left[a_{1+j} b_j + b_{1+j}\right] + b_{2+j}\right] + b_{3+j}}{1 - [a_0 a_1]^2}. \tag{7.58}$$

Figure 7.31 sketches (7.58) as a **zigzag-shaped** period-4 cycle when $x_0 = -4$, $a_0 = 1$, $a_1 = 2$, $b_0 = 1$, $b_1 = 1$, $b_2 = 1$ and $b_3 = 4$.

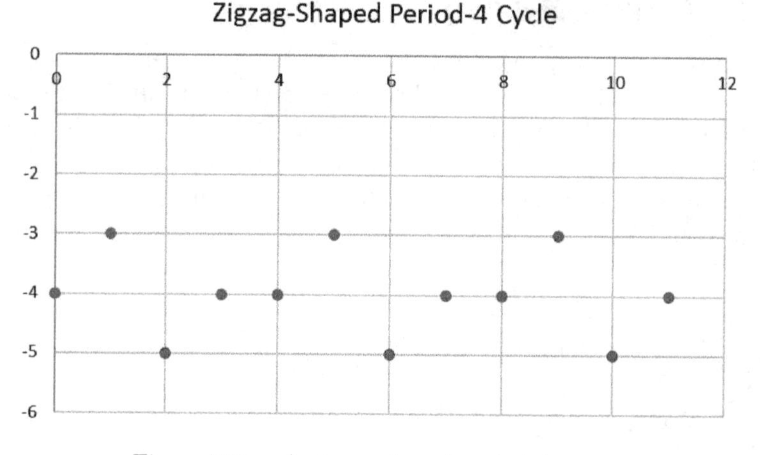

Figure 7.31: A zigzag-shaped period-4 cycle.

It is peculiar to note that when $\{a_n\}_{n=0}^{\infty}$ is periodic with period-2 and $\{b_n\}_{n=0}^{\infty}$ is periodic with period-4 you obtain a period-4 cycle as $lcm(2,4) = 4$.

Next, you will examine the periodic features of Eq. (7.1) when $\{a_n\}_{n=0}^{\infty}$ is periodic with period-2 and $\{b_n\}_{n=0}^{\infty}$ is periodic with period-3 and will show the existence of a unique period-6 cycle as $lcm(2,3) = 6$.

Set $x_6 = x_0$. Then by iterations you obtain the corresponding initial condition:

$$x_0 = \frac{a_0^2a_1^3b_0 + a_0^2a_1^2b_1 + a_0a_1^2b_2 + a_0a_1b_0 + a_1b_1 + b_2}{1 - [a_0a_1]^3}, \tag{7.59}$$

provided that $a_0a_1 \neq 1$. In the **sigma and product form** reformulate (7.59) as:

$$x_0 = \frac{\sum_{i=0}^{4}\left[\prod_{j=i+1}^{5} a_jb_i\right] + b_2}{1 - [a_0a_1]^3}, \tag{7.60}$$

In the **factored and nested form**, reformulate (7.59) as:

$$x_0 = \frac{a_1\left[a_0\left[a_1\left[a_0\left[a_1b_0 + b_1\right] + b_2\right] + b_0\right] + b_1\right] + b_2}{1 - [a_0a_1]^3}. \tag{7.61}$$

Then for all $j \in [0, 1, 2, 3, 4, 5]$, you can formulate the corresponding period-6 cycle of Eq. (7.1):

$$x_j = \frac{a_{1+j}\left[a_j\left[a_{1+j}\left[a_j\left[a_{1+j}b_j + b_{1+j}\right] + b_{2+j}\right] + b_j\right] + b_{1+j}\right] + b_{2+j}}{1 - [a_0 a_1]^3}. \tag{7.62}$$

Figure 7.32 traces (7.58) as a **scattered** period-6 cycle when $x_0 = 4$, $a_0 = -1$, $a_1 = 1$, $b_0 = 2$, $b_1 = 4$ and $b_2 = 2$.

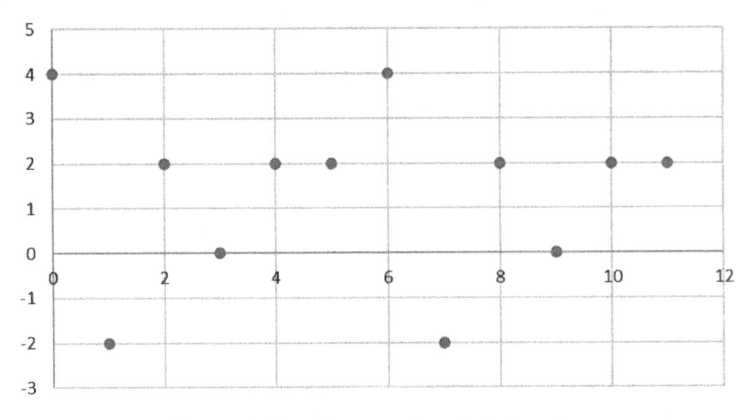

Figure 7.32: Scattered period-6 cycle.

Equations (7.60) and (7.61) then guide you to the following result.

Suppose that $\{a_n\}_{n=0}^{\infty}$ is periodic with period-k ($k \geq 2$) and $\{b_n\}_{n=0}^{\infty}$ is periodic with period-m ($m \geq 2$). In addition, assume that $k \neq m$ and let $L = lcm(k, m)$. Then Eq. (7.1) has a unique period-L cycle.

Proving the result will require decomposing into two cases when $k > m$ and when $k < m$.

7.5 Special Case of Eq. (7.1)

This section will adjourn the chapter with the periodic attributes of the following special case of Eq. (7.1):

$$x_{n+1} = -a_n x_n + 1, \quad n = 0, 1, \ldots, \tag{7.63}$$

where $\{a_n\}_{n=0}^{\infty}$ is a period-k sequence ($k \geq 2$). You will discover contrasting periodic patterns when $\{a_n\}_{n=0}^{\infty}$ is an even-ordered periodic sequence in comparison to when $\{a_n\}_{n=0}^{\infty}$ is an odd-ordered periodic sequence. First, assume that $\{a_n\}_{n=0}^{\infty}$ is a period-2 sequence and formulate the **explicit solution** of Eq. (7.63) and specify the periodicity character.

First, set $x_2 = x_0$ and by iterations you acquire:

$$\begin{aligned} &x_0, \\ &x_1 = -a_0 x_0 + 1, \\ &x_2 = -a_1 x_1 + 1 = a_0 a_1 x_0 - a_1 + 1 = x_0. \end{aligned} \tag{7.64}$$

Via (7.64), you obtain the corresponding initial condition:

$$x_0 = \frac{-a_1 + 1}{1 - a_0 a_1}, \tag{7.65}$$

provided that $a_0 a_1 \neq 1$ and acquire the corresponding unique period-2 cycle:

$$\frac{-a_1 + 1}{1 - a_0 a_1}, \frac{-a_0 + 1}{1 - a_0 a_1}, \ldots. \tag{7.66}$$

Figure 7.33 presents (7.66) as an **ascending and oscillatory period-2 cycle** with a negative term and a positive term when $x_0 = -0.2$, $a_0 = -3$ and $a_1 = 3$.

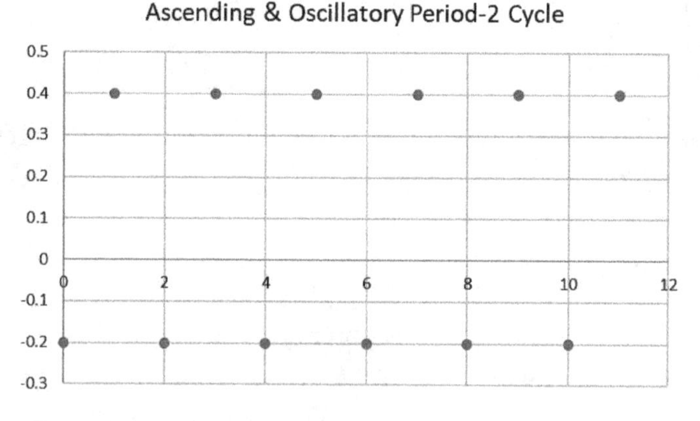

Figure 7.33: An ascending and oscillatory period-2 cycle.

Next, assume that $\{a_n\}_{n=0}^{\infty}$ is a period-3 sequence and formulate the **explicit solution** of Eq. (7.63) and specify the periodicity character.

We set $x_3 = x_0$ and by iterations you acquire the corresponding initial condition:

$$x_0 = \frac{a_1 a_2 - a_2 + 1}{1 + a_0 a_1 a_2}, \tag{7.67}$$

provided that $a_0 a_1 a_2 \neq -1$. Then for all $j \in [0, 1, 2]$, you formulate the corresponding period-3 cycle of Eq. (7.63):

$$x_j = \frac{a_{1+j} a_{2+j} - a_{2+j} + 1}{1 + a_0 a_1 a_2}. \tag{7.68}$$

Figure 7.34 traces (7.68) as an **ascending triangular-shaped** period-3 cycle with three positive terms when $x_0 = 0.2$, $a_0 = 1$, $a_1 = 1$ and $a_2 = 4$.

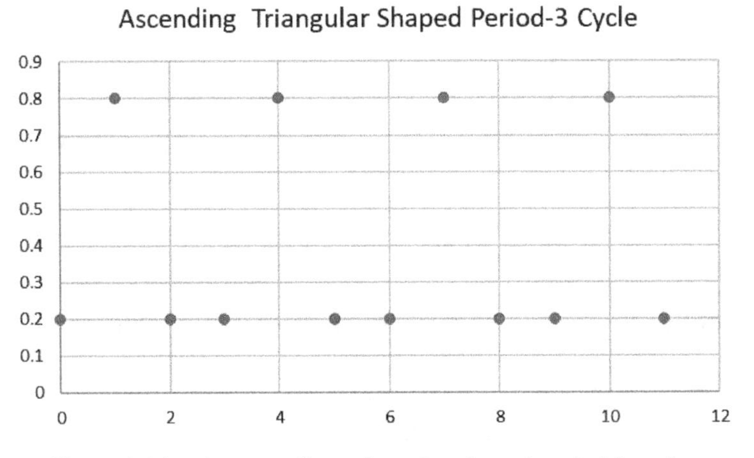

Figure 7.34: An ascending triangular-shaped period-3 cycle.

Next, you will outline the periodic traits of Eq. (7.63) when $\{a_n\}_{n=0}^{\infty}$ is a period-4 sequence and formulate the **explicit solution** of Eq. (7.63) and specify the periodicity character.

Set $x_4 = x_0$ and by iterations you acquire the corresponding initial condition:

$$x_0 = \frac{-a_1a_2a_3 + a_2a_3 - a_3 + 1}{1 - a_0a_1a_2a_3}, \tag{7.69}$$

where $a_0a_1a_2a_3 \neq 1$. Then for all $j \in [0, 1, 2, 3]$, you formulate the cognate period-4 cycle of Eq. (7.63):

$$x_j = \frac{-a_{1+j}a_{2+j}a_{3+j} + a_{2+j}a_{3+j} - a_{3+j} + 1}{1 - a_0a_1a_2a_3}. \tag{7.70}$$

Figure 7.35 renders (7.70) as an **oscillatory** period-4 cycle when $x_0 = 1.6$, $a_0 = 1$, $a_1 = -11$, $a_2 = 4$ and $a_3 = 1$.

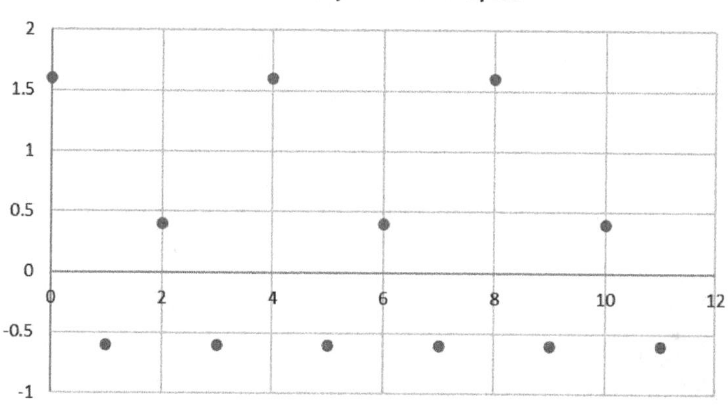

Figure 7.35: An oscillatory period-4 cycle.

Note that in the **sigma and product form** we reformulate (7.69) as:

$$x_0 = \frac{\sum_{i=0}^{2} \left[\prod_{j=i+1}^{3} (-1)^{i+1} a_j\right] + 1}{1 - [a_0a_1a_2a_3]}. \tag{7.71}$$

Equations (7.65), (7.67) and (7.69) guide you to the following results when $\{a_n\}_{n=0}^{\infty}$ is an even-ordered and an odd-ordered periodic sequence.

First suppose that $\{a_n\}_{n=0}^{\infty}$ is periodic with period-$2k$ ($k \in \mathbb{N}$). Then Eq. (7.63) has a **unique period-$2k$ cycle** ($k \in \mathbb{N}$) with the corresponding initial condition:

$$x_0 = \frac{\sum_{i=0}^{2k-2} \left[\prod_{j=i+1}^{2k-1} (-1)^{i+1} a_j\right] + 1}{1 - [\prod_{i=0}^{2k-1} a_i]}, \tag{7.72}$$

where

$$\prod_{i=0}^{2k-1} a_i \neq 1.$$

Next, suppose that $\{a_n\}_{n=0}^{\infty}$ is periodic with period-$(2k+1)$ ($k \in \mathbb{N}$). Then Eq. (7.63) has a **unique period-$(2k+1)$ cycle** ($k \in \mathbb{N}$) with the corresponding initial condition:

$$x_0 = \frac{\sum_{i=0}^{2k-1} \left[\prod_{j=i+1}^{2k} (-1)^{i} a_j\right] + 1}{1 + [\prod_{i=0}^{2k} a_i]}, \tag{7.73}$$

where

$$\prod_{i=0}^{2k} a_i \neq -1.$$

7.6 Exercises

In problems 1–8, determine the periodic traits by solving the given **initial value problem**:

1. $\begin{cases} x_{n+1} = -\, x_n, & n = 0, 1, \ldots. \\ x_0 = -8. \end{cases}$

2. $\begin{cases} x_{n+1} = -\, x_n, & n = 0, 1, \ldots. \\ x_0 = 6. \end{cases}$

3. $\begin{cases} x_{n+1} = -\, x_n + 2, & n = 0, 1, \ldots. \\ x_0 = -6. \end{cases}$

4. $\begin{cases} x_{n+1} = -\, x_n + 4, & n = 0, 1, \ldots. \\ x_0 = 4. \end{cases}$

5. $\begin{cases} x_{n+1} = (-1)^{n+1}\, x_n, & n = 0, 1, \ldots. \\ x_0 = 2. \end{cases}$

6. $\begin{cases} x_{n+1} = (-1)^{n}\, x_n, & n = 0, 1, \ldots. \\ x_0 = -1. \end{cases}$

7. $\begin{cases} x_{n+1} = (-1)^{n}\, x_n + 1, & n = 0, 1, \ldots. \\ x_0 = 3. \end{cases}$

8. $\begin{cases} x_{n+1} = (-1)^{n}\, x_n - 1, & n = 0, 1, \ldots. \\ x_0 = -2. \end{cases}$

In problems 9–12, solve the following recursive relation:

$$x_{n+1} = a_n x_n, \quad n = 0, 1, \ldots,$$

and determine the periodic character when:

9. $\{a_n\}_{n=0}^{\infty}$ is a period-2 sequence and $a_0 a_1 = -1$.
10. $\{a_n\}_{n=0}^{\infty}$ is a period-3 sequence and $a_0 a_1 a_2 = -1$.
11. $\{a_n\}_{n=0}^{\infty}$ is a period-4 sequence and $a_0 a_1 a_2 a_3 = -1$.
12. $\{a_n\}_{n=0}^{\infty}$ is a period-k sequence ($k \geq 2$) and $\prod_{i=0}^{k-1} a_i = -1$.

In problems 13–16, determine the periodic traits of:

$$x_{n+1} = a_n x_n + 1, \quad n = 0, 1, \ldots,$$

13. When $\{a_n\}_{n=0}^{\infty}$ is a period-2 sequence (period-2 cycle).
14. When $\{a_n\}_{n=0}^{\infty}$ is a period-3 sequence (period-3 cycle).
15. When $\{a_n\}_{n=0}^{\infty}$ is a period-4 sequence (period-4 cycle).
16. When $\{a_n\}_{n=0}^{\infty}$ is a period-k sequence, ($k \geq 2$) (period-k cycle).

In problems 17–20, determine the periodic features of:

$$x_{n+1} = -a_n x_n + 1, \quad n = 0, 1, \ldots,$$

17. When $\{a_n\}_{n=0}^{\infty}$ is a period-5 sequence (period-5 cycle).
18. When $\{a_n\}_{n=0}^{\infty}$ is a period-$(2k+1)$ sequence, $(k \geq 1)$ (period-$(2k+1)$ cycle).
19. When $\{a_n\}_{n=0}^{\infty}$ is a period-6 sequence (period-4 cycle).
20. When $\{a_n\}_{n=0}^{\infty}$ is a period-$2k$ sequence, $(k \geq 1)$ (period-$2k$ cycle).

In problems 21–24, determine the periodicity attributes of:

$$x_{n+1} = -x_n + b_n, \quad n = 0, 1, \ldots,$$

21. When $\{b_n\}_{n=0}^{\infty}$ is a period-5 sequence (period-5 cycle).
22. When $\{b_n\}_{n=0}^{\infty}$ is a period-7 sequence (period-7 cycle).
23. When $\{b_n\}_{n=0}^{\infty}$ is a period-9 sequence (period-9 cycle).
24. When $\{b_n\}_{n=0}^{\infty}$ is a period-$(2k+1)$ sequence, $(k \geq 1)$ (period-$(2k+1)$ cycle).

In problems 25–28, determine the periodic features of:

$$x_{n+1} = -x_n + b_n, \quad n = 0, 1, \ldots,$$

25. When $\{b_n\}_{n=0}^{\infty}$ is a period-6 sequence (period-6 cycles).
26. When $\{b_n\}_{n=0}^{\infty}$ is a period-8 sequence (period-8 cycles).
27. When $\{b_n\}_{n=0}^{\infty}$ is a period-10 sequence (period-10 cycles).
28. When $\{b_n\}_{n=0}^{\infty}$ is a period-$2k$ sequence, $(k \geq 2)$ (period-$2k$ cycles).

In problems 29–38, determine the periodic attributes of:

$$x_{n+1} = a_n x_n + b_n, \quad n = 0, 1, \ldots,$$

29. When $\{a_n\}_{n=0}^{\infty}$ and $\{b_n\}_{n=0}^{\infty}$ are period-4 sequences (period-4 cycle).
30. When $\{a_n\}_{n=0}^{\infty}$ and $\{b_n\}_{n=0}^{\infty}$ are period-5 sequences (period-5 cycle).
31. When $\{a_n\}_{n=0}^{\infty}$ and $\{b_n\}_{n=0}^{\infty}$ are period-6 sequences (period-6 cycle).

32. When $\{a_n\}_{n=0}^{\infty}$ and $\{b_n\}_{n=0}^{\infty}$ are period-k sequences, $(k \geq 2)$ (period-k cycle).
33. When $\{a_n\}_{n=0}^{\infty}$ is a period-4 sequence and $\{b_n\}_{n=0}^{\infty}$ is a period-2 sequence (period-4 cycle).
34. When $\{a_n\}_{n=0}^{\infty}$ is a period-3 sequence and $\{b_n\}_{n=0}^{\infty}$ is a period-6 sequence (period-6 cycle).
35. When $\{a_n\}_{n=0}^{\infty}$ is a period-6 sequence and $\{b_n\}_{n=0}^{\infty}$ is a period-3 sequence (period-6 cycle).
36. When $\{a_n\}_{n=0}^{\infty}$ is a period-4 sequence and $\{b_n\}_{n=0}^{\infty}$ is a period-8 sequence (period-8 cycle).
37. When $\{a_n\}_{n=0}^{\infty}$ is a period-k sequence and $\{b_n\}_{n=0}^{\infty}$ is a period-m sequence where $lcm(k, m) = 1$ and $k < m$ (period-km cycle).
38. When $\{a_n\}_{n=0}^{\infty}$ is a period-k sequence and $\{b_n\}_{n=0}^{\infty}$ is a period-m sequence where $lcm(k, m) = 1$ and $k > m$ (period-km cycle).

Chapter 8

Answers to Chapter Exercises

8.1 Answers to Chapter 1 Exercises

1. $\{6n\}_{n=1}^{\infty}$.
3. $\{4(2n-1)\}_{n=1}^{\infty}$.
5. $\{5n\}_{n=6}^{\infty}$.
7. $\{3n+1\}_{n=1}^{\infty}$.
9. $\{7n-3\}_{n=1}^{\infty}$.
11. $\{8 \cdot 2^n\}_{n=0}^{\infty} = \{2^{n+3}\}_{n=0}^{\infty}$.
13. $\{\frac{3^n}{9}\}_{n=0}^{\infty} = \{3^{n-2}\}_{n=0}^{\infty}$.
15. $\{5 \cdot 3^n\}_{n=0}^{\infty}$.
17. $\{64 \cdot \left(\frac{3}{4}\right)^n\}_{n=0}^{\infty}$.
19. For all $n \geq 0$:

$$x_n = \begin{cases} 2 & \text{if } n \text{ is even,} \\ -4 & \text{if } n \text{ is odd.} \end{cases}$$

21. For all $n \geq 0$:

$$x_n = \begin{cases} 3 & \text{if } n = 3\text{k}, \\ 1 & \text{if } n = 3\text{k} + 1, \\ -1 & \text{if } n = 3\text{k} + 2. \end{cases}$$

23. For all $n \geq 0$:

$$x_n = \begin{cases} 2n+1 & \text{if } n \text{ is even,} \\ -2[n+1] & \text{if } n \text{ is odd.} \end{cases}$$

25. For all $n \geq 0$:

$$x_n = \begin{cases} 2^n & \text{if } n \text{ is even,} \\ -2^n & \text{if } n \text{ is odd.} \end{cases}$$

27. For all $n \geq 0$:

$$y = \begin{cases} -1 & \text{if } \ x \in [0,2] \\ 1 & \text{if } \ x \in [2,4] \\ 3 & \text{if } \ x \in [4,6] \\ 5 & \text{if } \ x \in [6,8] \\ \vdots & \\ 2n-1 & \text{if } \ x \in [2n, 2n+2] \\ \vdots & \end{cases}$$

29. For all $n \geq 0$:

$$y = \begin{cases} 2-x & \text{if } \ x \in [0,2] \\ x-2 & \text{if } \ x \in [2,4] \\ 6-x & \text{if } \ x \in [4,6] \\ x-6 & \text{if } \ x \in [6,8] \\ \vdots & \\ (4n+2)-x & \text{if } \ x \in [2n, 2n+2] \\ x-(4n+2) & \text{if } \ x \in [2n+2, 2n+4] \\ \vdots & \end{cases}$$

31. $y = 0, \ x \in [0,8]$,
$y = 8 - x, \ x \in [0,8]$,
$x = 0, \ y \in [0,8]$.

8.2 Answers to Chapter 2 Exercises

1. For all $n \geq 0$:

$$y = \begin{cases} 5, & x \in [3n, 3n+1] \\ 3, & x \in [3n+1, 3n+2] \\ 1, & x \in [3n+2, 3n+3] \end{cases}$$

3. For all $n \geq 0$:

$$y = \begin{cases} 1, & x \in [3n, 3n+1] \\ x - n, & x \in [3n+1, 3n+3] \end{cases}$$

5. For all $n \geq 0$:

$$y = \begin{cases} x - 6n, & x \in [6n, 6n+2] \\ (4+6n) - x, & x \in [6n+2, 6n+3] \\ x - (2+6n), & x \in [6n+3, 6n+4] \\ (6+6n) - x, & x \in [6n+4, 6n+6] \end{cases}$$

7. $y = \begin{cases} 4, & x \in [-4, 4] \\ 0, & x \in [0, 4] \\ -4, & x \in [-4, 4] \end{cases}$ $\quad x = \begin{cases} 4, & y \in [-4, 4] \\ 0, & y \in [-4, 4] \\ -4, & y \in [-4, 4] \end{cases}$

9. $y = \begin{cases} 4, & x \in [-4, 4] \\ 0, & x \in [0, 4] \\ -4, & x \in [-4, 4] \\ -2, & x \in [2, 4] \end{cases}$ $\quad x = \begin{cases} 4, & y \in [-4, 4] \\ 0, & y \in [-4, 4] \\ -4, & y \in [-4, 4] \\ 2, & y \in [-4, 0] \end{cases}$

11. $y = \begin{cases} 4, & x \in [-4, 4] \\ 0, & x \in [0, 4] \\ -4, & x \in [-4, 4] \\ -2, & x \in [2, 4] \\ -3, & x \in [3, 4] \end{cases}$ $\quad x = \begin{cases} 4, & y \in [-4, 4] \\ 0, & y \in [-4, 4] \\ -4, & y \in [-4, 4] \\ 2, & y \in [2, 4] \\ 3, & y \in [-4, -2] \end{cases}$

13. $y = \begin{cases} 0, & x \in [0,2] \\ 1, & x \in [0,1] \end{cases}$ $\quad x = \begin{cases} 0, & y \in [0,2] \\ 1, & y \in [0,1] \end{cases}$

$$y = \begin{cases} 2 - x, & x \in [0,2] \\ 1 - x, & x \in [0,1] \end{cases}$$

15. $y = \begin{cases} 0, & x \in [0,4] \\ 1, & x \in [0,3] \\ 2, & x \in [0,2] \\ 3, & x \in [0,1] \end{cases}$ $\quad x = \begin{cases} 0, & y \in [0,4] \\ 1, & y \in [0,3] \\ 2, & y \in [0,2] \\ 3, & y \in [0,1] \end{cases}$

$$y = \begin{cases} 4 - x, & x \in [0,4] \\ 3 - x, & x \in [0,3] \\ 2 - x, & x \in [0,2] \\ 1 - x, & x \in [0,1] \end{cases}$$

17. $y = \begin{cases} 4, & x \in [-4,4] \\ 8, & x \in [-8,8] \end{cases}$ $\quad y = \begin{cases} x, & x \in [0,8] \\ x + 8, & x \in [-4,0] \\ \\ -x, & x \in [8,0] \\ -x + 8, & x \in [0,4] \end{cases}$

19. $y = \begin{cases} 8, & x \in [-8,8] \\ 6, & x \in [-2,2] \\ 4, & x \in [-4,4] \end{cases}$ $\quad y = \begin{cases} x, & x \in [0,8] \\ x + 4, & x \in [0,2] \\ x + 8, & x \in [-4,0] \\ \\ -x, & x \in [-8,0] \\ -x + 4, & x \in [-2,0] \\ -x + 8, & x \in [0,4] \end{cases}$

21. $$x = \begin{cases} 4, & y \in [-4,4] \\ -4, & y \in [-4,4] \end{cases} \qquad y = \begin{cases} 4, & x \in [-4,4] \\ -4, & x \in [-4,4] \\ \\ x-4, & x \in [0,4] \\ x+4, & x \in [-4,0] \\ -x+4, & x \in [0,4] \\ -x-4, & x \in [-4,0] \end{cases}$$

23. $$x = \begin{cases} 4, & y \in [-4,4] \\ -4, & y \in [-4,4] \\ 2, & y \in [-2,2] \\ -2, & y \in [-2,2] \end{cases} \qquad y = \begin{cases} 4, & x \in [-4,4] \\ -4, & x \in [-4,4] \\ 2, & x \in [-2,2] \\ -2, & x \in [-2,2] \\ \\ x-4, & x \in [0,4] \\ x+4, & x \in [-4,0] \\ -x+4, & x \in [0,4] \\ -x-4, & x \in [-4,0] \\ \\ x-2, & x \in [0,2] \\ x+2, & x \in [-2,0] \\ -x+2, & x \in [0,2] \\ -x-2, & x \in [-2,0] \end{cases}$$

25. $$y = \begin{cases} 0, & x \in [0,16] \\ 4, & x \in [0,4] \\ 8, & x \in [0,8] \end{cases} \qquad x = \begin{cases} 0, & y \in [0,16] \\ 4, & y \in [0,4] \\ 8, & y \in [0,8] \end{cases}$$

$$y = \begin{cases} 4-x, & x \in [0,4] \\ 8-x, & x \in [0,8] \\ 16-x, & x \in [0,16] \end{cases}$$

27. $$y = \begin{cases} 0, & x \in [0,16] \\ 1, & x \in [0,1] \\ 2, & x \in [0,2] \\ 4, & x \in [0,4] \\ 8, & x \in [0,8] \end{cases} \qquad x = \begin{cases} 0, & y \in [0,16] \\ 1, & y \in [0,1] \\ 2, & y \in [0,2] \\ 4, & y \in [0,4] \\ 8, & y \in [0,8] \end{cases}$$

$$y = \begin{cases} 1 - x, & x \in [0,1] \\ 2 - x, & x \in [0,2] \\ 4 - x, & x \in [0,4] \\ 8 - x, & x \in [0,8] \\ 16 - x, & x \in [0,16] \end{cases}$$

29. $$y = \begin{cases} 8, & x \in [-8,8] \\ 6, & x \in [-8,-6] \\ 4, & x \in [-8,-4] \\ 2, & x \in [-8,-2] \\ 0, & x \in [-8,0] \\ -4, & x \in [-8,0] \\ -8, & x \in [-8,8] \end{cases} \qquad x = \begin{cases} 8, & y \in [-8,8] \\ 0, & y \in [-8,0] \\ -2, & y \in [0,2] \\ -4, & y \in [-8,4] \\ 0, & y \in [-8,0] \\ -6, & y \in [0,6] \\ -8, & y \in [-8,8] \end{cases}$$

$$y = \begin{cases} x - 8, & x \in [0,4] \\ x - 0, & x \in [-8,0] \\ x + 4, & x \in [-8,-2] \\ x + 8, & x \in [-8,-4] \\ x + 12, & x \in [-8,-6] \\ -x + 0, & x \in [-8,-8] \\ -x - 4, & x \in [-8,-2] \\ -x - 8, & x \in [-8,0] \\ -x - 12, & x \in [-8,-6] \end{cases}$$

31. $\{3i+1\}_{i=0}^{n}$.

33. $k = 1$.

8.3 Answers to Chapter 3 Exercises

1. $\{8n-1\}_{n=1}^{\infty}$.
3. $\{7(2n-1)\}_{n=1}^{\infty}$.
5. $\{(2n-1)^2\}_{n=1}^{\infty}$.
7. $\{(4n-1)^2\}_{n=1}^{\infty}$.
9. $\{(4n-2)^2\}_{n=1}^{\infty}$.
11. For all $n \geq 0$:

$$\{x_n\}_{n=0}^{\infty} = \begin{cases} 4 & \text{if } n = 0, \\ 4 + [\sum_{i=i}^{n} (2i+1)] & \text{if } n \geq 1. \end{cases}$$

13. $\{2 \cdot 3^n\}_{n=0}^{\infty}$.
15. $\{2 \cdot [\sqrt{3}]^n\}_{n=0}^{\infty}$.
17. $\{12 \cdot \left(\frac{3}{2}\right)^n\}_{n=0}^{\infty}$.
19. $\prod_{i=i}^{n} (2i-1)$.
21. $\prod_{i=i}^{n} (4i-1)$.
23. For all $n \geq 0$:

$$\{x_n\}_{n=0}^{\infty} = \begin{cases} 5 & \text{if } n = 0, \\ 5 \cdot [\prod_{i=i}^{n} (i+3)] & \text{if } n \geq 1. \end{cases}$$

25. For all $n \geq 0$:

$$\{x_n\}_{n=0}^{\infty} = \begin{cases} 1 & \text{if } n = 0, \\ \prod_{i=i}^{n} 2i & \text{if } n \geq 1. \end{cases}$$

27. For all $n \geq 0$:

$$\{x_n\}_{n=0}^{\infty} = \begin{cases} -[6n+4] & \text{if } n \text{ is even}, \\ [6n+4] & \text{if } n \text{ is odd}. \end{cases}$$

29. For all $n \geq 0$:

$$\{x_n\}_{n=0}^{\infty} = \begin{cases} (-1)^{\frac{n+2}{2}} 2^{\frac{n}{2}} & \text{if } n \text{ is even}, \\ [\sqrt{2}]^n & \text{if } n \text{ is odd}. \end{cases}$$

31. For all $n \geq 0$:

$$\{x_n\}_{n=0}^{\infty} = \begin{cases} (-1)^{\frac{n}{2}} 5[n+1] & \text{if } n \text{ is even,} \\ (-1)^{\frac{n-1}{2}} 5[n+1] & \text{if } n \text{ is odd.} \end{cases}$$

33. $2^{n \cdot [n+1]}$, for all $n \geq 1$.

35. $\frac{2}{n \cdot [n-1]}$, for all $n \geq 3$.

8.4 Answers to Chapter 4 Exercises

1. $3 \cdot 40 \cdot 81$.

3. $20 + 40 \cdot 21$.

5. $\frac{2^{10}-1}{4}$.

7. $\frac{3^9+1}{36}$.

9. $\frac{[k-6] \cdot [k+7]}{2}$.

8.5 Answers to Chapter 5 Exercises

1. 84.

3. 90.

5. $\binom{7}{3}$.

7. $[k+2] \cdot [k+1]$.

9. $\prod_{i=1}^{n} [k+i]$.

11. $8 \cdot 7 \cdot 6 \cdot 5 = \prod_{i=0}^{3} [8-i]$.

13. $\prod_{i=0}^{k-1} [2k-i]$.

15. $\binom{n+1}{k}$.

17. $x^2 + y^2 + z^2 + 2[xy + xz + yz]$.

19. $\sum_{i=0}^{6} \binom{6}{i} a^{18-3i} b^{3i}$.

21. $\sum_{i=0}^{12} \binom{12}{i} x^{12-3i}$.

23. $\binom{18}{5}$.

25. $\binom{2n}{n}$.

8.6 Answers to Chapter 6 Exercises

1. For all $n \geq 0$:

$$\begin{cases} x_{n+1} = x_n + 5, \\ x_0 = 2. \end{cases}$$

3. For all $n \geq 0$:

$$\begin{cases} x_{n+1} = x_n + 2(n+1), \\ x_0 = 5. \end{cases}$$

5. For all $n \geq 0$:

$$\begin{cases} x_{n+1} = x_n + 4(n+1), \\ x_0 = 3. \end{cases}$$

7. For all $n \geq 0$:

$$\begin{cases} x_{n+1} = x_n + 3(2n+1), \\ x_0 = 5. \end{cases}$$

9. For all $n \geq 0$:

$$\begin{cases} x_{n+1} = 3x_n, \\ x_0 = 4. \end{cases}$$

11. For all $n \geq 0$:

$$\begin{cases} x_{n+1} = \frac{2}{3}x_n, \\ x_0 = 54. \end{cases}$$

13. For all $n \geq 0$:

$$\begin{cases} x_{n+1} = 2(n+1)x_n, \\ x_0 = 1. \end{cases}$$

15. For all $n \geq 0$:

$$\begin{cases} x_{n+1} = (4n+5)x_n, \\ x_0 = 1. \end{cases}$$

17. For all $n \geq 0$:

$$\begin{cases} x_{n+1} = \frac{[2n+4]\cdot[2n+6]\cdot x_n}{2n+2}, \\ x_0 = 3. \end{cases}$$

19. For all $n \geq 0$:

$$\begin{cases} x_{n+1} = x_n + (2n+3), \\ x_0 = 1. \end{cases}$$

21. For all $n \geq 0$:

$$\begin{cases} x_{n+1} = x_n + (n+1)^2, \\ x_0 = 1. \end{cases}$$

29. For all $n \geq 0$, $x_n = \left(\frac{1}{2}\right)^{n+3}$.

31. For all $n \geq 0$, $x_n = \left(\frac{4}{3}\right)^{n+2}$.

33. For all $n \geq 0$, $x_n = \left(\frac{3}{7}\right)^n + 3$.

35. For all $n \geq 0$, $x_n = \frac{3-n}{2}$.

37. For all $n \geq 0$:

$$x_n = \begin{cases} -2 & \text{if } n \text{ is even}, \\ 9 & \text{if } n \text{ is odd}. \end{cases}$$

39. For all $n \geq 0$, $x_n = (-1)^n[n+4]$.

41. For all $n \geq 0$, $x_n = \sum_{i=0}^{n} 2^{i+2} = 4[2^{n+1} - 1]$.

43. For all $n \geq 0$, $x_n = \sum_{i=0}^{n} (i+1)^2 = \frac{n[n+1][2n+1]}{6}$.

8.7 Answers to Chapter 7 Exercises

1. For all $n \geq 0$:

$$x_n = \begin{cases} -8 & \text{if } n \text{ is even,} \\ 8 & \text{if } n \text{ is odd.} \end{cases}$$

3. For all $n \geq 0$:

$$x_n = \begin{cases} -6 & \text{if } n \text{ is even,} \\ 8 & \text{if } n \text{ is odd.} \end{cases}$$

5. For all $n \geq 0$:

$$x_n = \begin{cases} 2 & \text{if } n = 4k, \\ -2 & \text{if } n = 4k+1, \\ -2 & \text{if } n = 4k+2, \\ 2 & \text{if } n = 4k+3. \end{cases}$$

7. For all $n \geq 0$:

$$x_n = \begin{cases} 3 & \text{if } n = 4k, \\ 4 & \text{if } n = 4k+1, \\ -3 & \text{if } n = 4k+2, \\ -2 & \text{if } n = 4k+3. \end{cases}$$

9. Alternating period-4 pattern:

$$x_0, \ a_0x_0, \ -x_0, \ a_0x_0, \ \ldots.$$

11. Alternating period-8 pattern:

$$x_0, \ a_0x_0, \ a_0a_1x_0, \ a_0a_1a_2x_0, \ -x_0, \ -a_0x_0, \ -a_0a_1x_0,$$
$$-a_0a_1a_2x_0, \ \ldots.$$

13. Unique period-2 pattern:

$$\frac{a_1+1}{1-a_0a_1}, \ \frac{a_0+1}{1-a_0a_1}, \ \ldots.$$

15. Unique period-4 pattern. For $i = 0, 1, 2, 3$:

$$\frac{a_{1+i}a_{2+i}a_{3+i} + a_{2+i}a_{3+i} + a_{3+i} + 1}{1 - a_0a_1a_2a_3}.$$

17. Unique period-4 pattern. For $i = 0, 1, 2, 3$:

$$\frac{-a_{1+i}a_{2+i}a_{3+i} + a_{2+i}a_{3+i} - a_{3+i} + 1}{1 - a_0a_1a_2a_3}.$$

19. Unique period-5 pattern. For $i = 0, 1, 2, 3, 4$:

$$\frac{a_{1+i}a_{2+i}a_{3+i}a_{4+i} - a_{2+i}a_{3+i}a_{4+i} + a_{3+i}a_{4+i} - a_{4+i} + 1}{1 + a_0a_1a_2a_3a_4}.$$

21. Unique period-5 pattern with:

$$x_0 = \frac{b_0 - b_1 + b_2 - b_3 + b_4}{2} = \frac{\sum_{i=0}^{2} b_{2i} - \sum_{i=0}^{1} b_{2i+1}}{2}.$$

23. Unique period-9 pattern with:

$$x_0 = \frac{\sum_{i=0}^{4} b_{2i} - \sum_{i=0}^{3} b_{2i+1}}{2}.$$

25. Provided that $\sum_{i=0}^{2} b_{2i} = \sum_{i=0}^{2} b_{2i+1}$, then for all $i \in \{0, 1, 2, 3, 4, 5\}$ we obtain a period-6 pattern:

$$x_i = (-1)^i x_0 + (-1)^{i+1} \sum_{j=0}^{i} (-1)^j b_j.$$

27. Provided that $\sum_{i=0}^{4} b_{2i} = \sum_{i=0}^{4} b_{2i+1}$, then for all $i \in \{0, 1, \ldots, 9\}$ we obtain a period-10 pattern:

$$x_i = (-1)^i x_0 + (-1)^{i+1} \sum_{j=0}^{i} (-1)^j b_j.$$

29. Unique period-4 pattern with:

$$x_0 = \frac{a_3\left[a_2\left[a_1b_0 + b_1\right] + b_2\right] + b_3}{1 - a_0a_1a_2a_3} = \frac{\sum_{i=0}^{2}\left[\prod_{j=i+1}^{3} a_jb_i\right] + b_3}{1 - a_0a_1a_2a_3}.$$

31. Unique period-6 pattern with:

$$x_0 = \frac{a_5\left[a_4\left[a_3\left[a_2\left[a_1 b_0 + b_1\right] + b_2\right] + b_3\right] + b_4\right] + b_5}{1 - a_0 a_1 a_2 a_3 a_4 a_5}$$

$$= \frac{\sum_{i=0}^{4}\left[\prod_{j=i+1}^{5} a_j b_i\right] + b_5}{1 - a_0 a_1 a_2 a_3 a_4 a_5}.$$

33. Unique period-4 pattern with:

$$x_0 = \frac{a_1\left[a_0\left[a_1 b_0 + b_1\right] + b_2\right] + b_3}{1 - \left[a_0 a_1\right]^2} = \frac{\sum_{i=0}^{2}\left[\prod_{j=i+1}^{3} a_j b_i\right] + b_3}{1 - \left[a_0 a_1\right]^2}.$$

35. Unique period-6 pattern with:

$$x_0 = \frac{a_2\left[a_1\left[a_0\left[a_2\left[a_1 b_0 + b_1\right] + b_2\right] + b_3\right] + b_4\right] + b_5}{1 - \left[a_0 a_1 a_2\right]^2}$$

$$= \frac{\sum_{i=0}^{4}\left[\prod_{j=i+1}^{5} a_j b_i\right] + b_5}{1 - \left[a_0 a_1 a_2\right]^2}.$$

Chapter 9

Appendices

9.1 Right Triangles

1. Pythagorean Theorem:

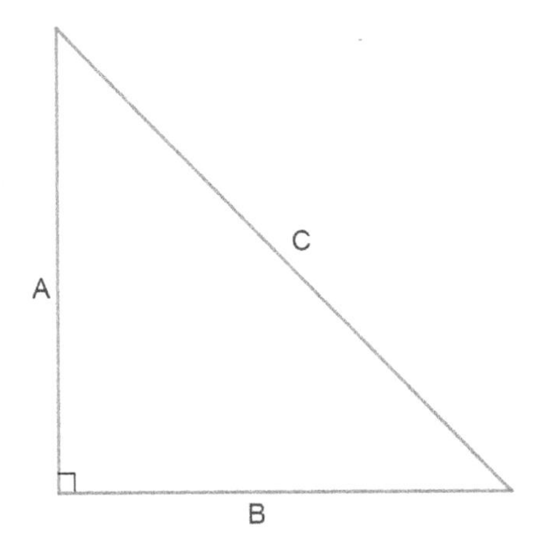

$$A^2 + B^2 = C^2.$$

2. 3-4-5 Triangle:

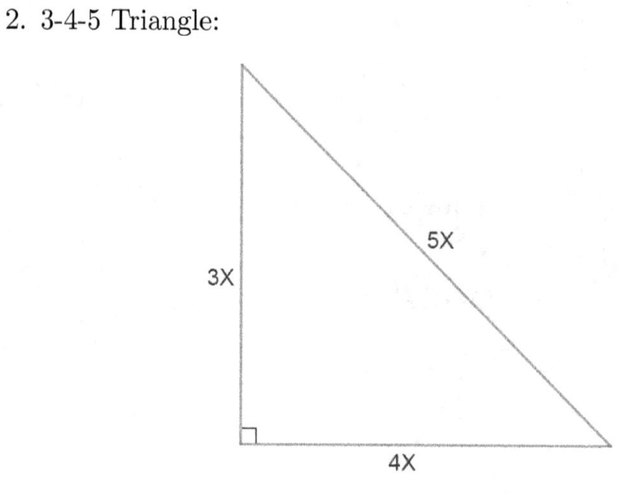

3. 5-12-13 Triangle:

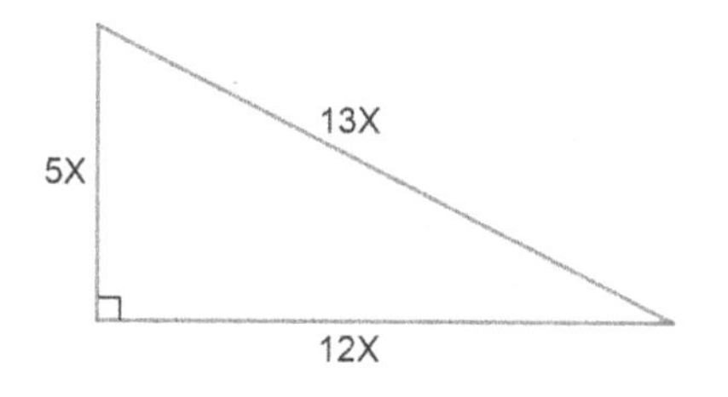

4. 45-45-90 Triangle:

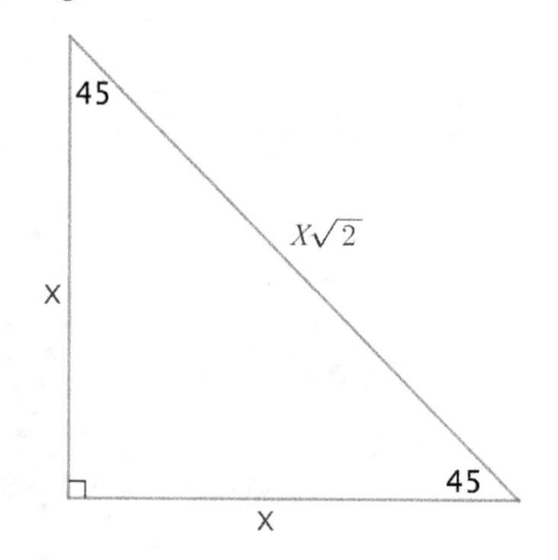

5. 30-60-90 Triangle:

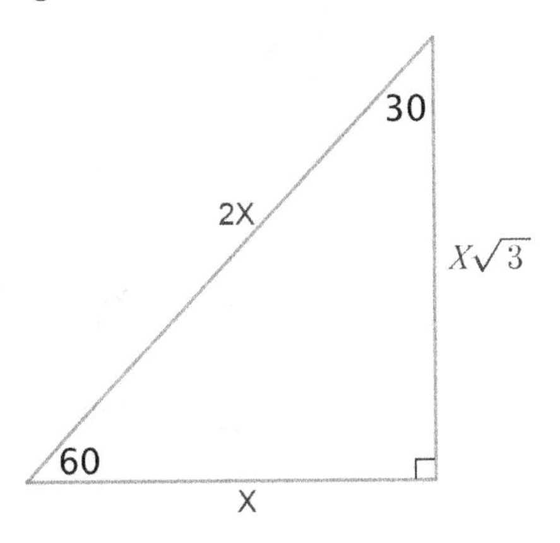

9.2 Isosceles Triangle:

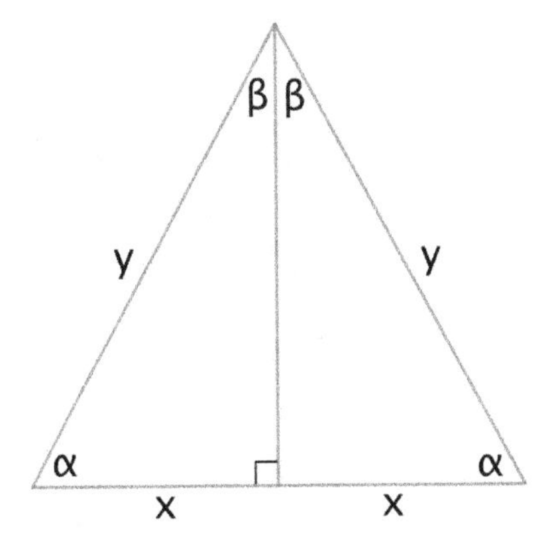

9.3 Equilateral Triangle:

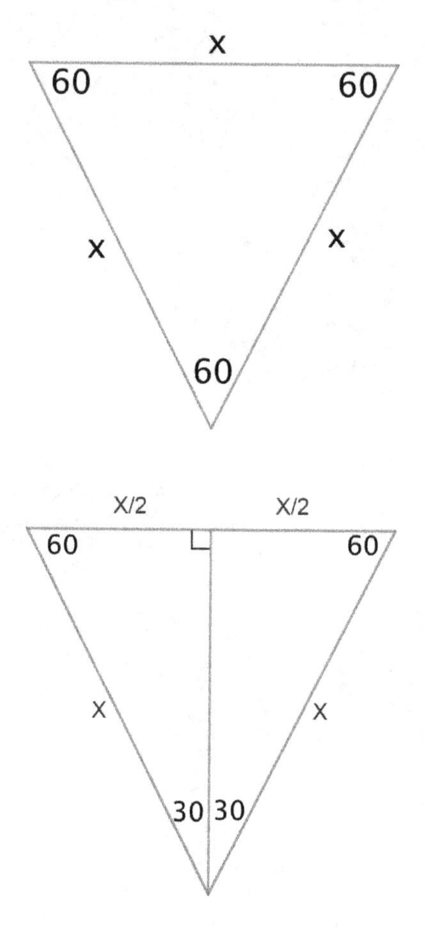

9.4 Area of Figures

1. Area of a circle:

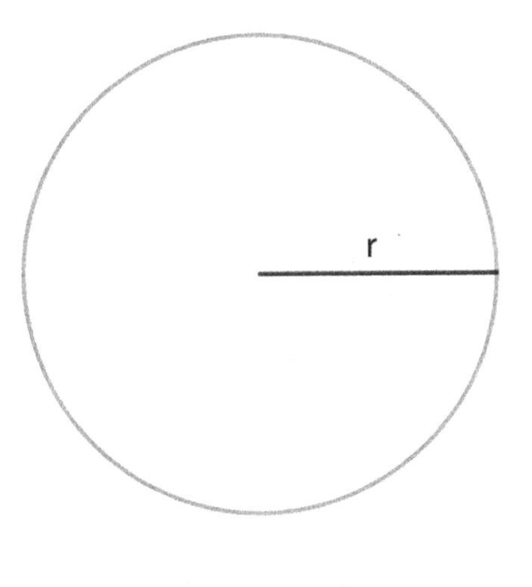

$$A = \pi\, r^2.$$

2. Area of a square:

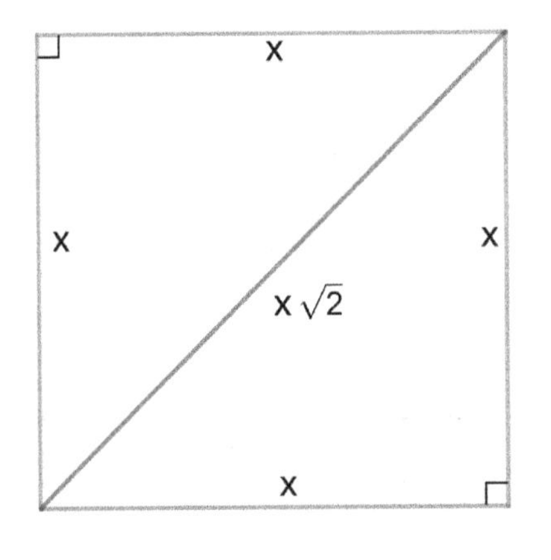

$$A = x^2.$$

3. Area of a rectangle:

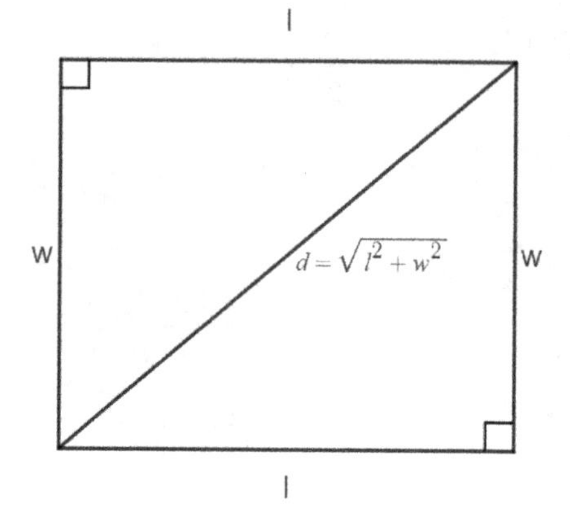

$$A = l \cdot w.$$

4. Area of a triangle:

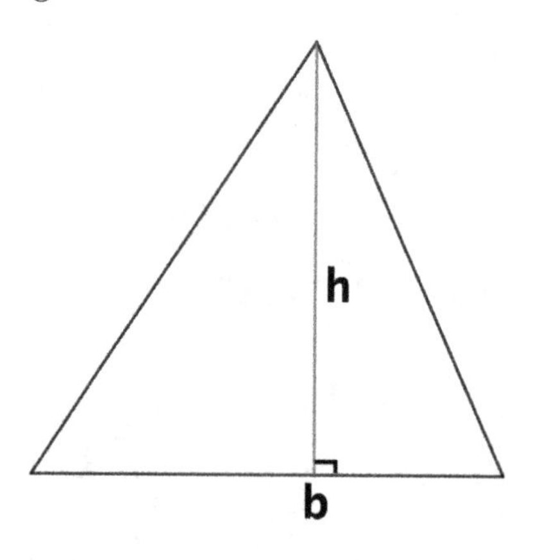

$$A = \frac{b \cdot h}{2}.$$

9.5 Patterns (Sequences)

1. Linear patterns:

$$1,\ 2,\ 3,\ 4,\ 5,\ 6,\ 7,\ \ldots = \{n\}_{n=1}^{\infty}$$

$$2,\ 4,\ 6,\ 8,\ 10,\ 12,\ 14,\ \ldots = \{2n\}_{n=1}^{\infty}$$

$$1,\ 3,\ 5,\ 7,\ 9,\ 11,\ 13,\ \ldots = \{2n+1\}_{n=0}^{\infty}$$

$$3,\ 6,\ 9,\ 12,\ 15,\ 18,\ 21,\ \ldots = \{3n\}_{n=1}^{\infty}$$

$$4,\ 8,\ 12,\ 16,\ 20,\ 24,\ 28,\ \ldots = \{4n\}_{n=1}^{\infty}.$$

2. Quadratic patterns:

$$1,\ 4,\ 9,\ 16,\ 25,\ 36,\ 49,\ \ldots = \{n^2\}_{n=1}^{\infty}$$

$$4,\ 16,\ 36,\ 64,\ 100,\ 144,\ 196,\ \ldots = \{(2n)^2\}_{n=1}^{\infty}$$

$$1,\ 9,\ 25,\ 49,\ 81,\ 121,\ 169,\ \ldots = \{(2n-1)^2\}_{n=1}^{\infty}$$

$$1,\ 25,\ 81,\ 169,\ 289,\ 441,\ 625,\ \ldots = \{(4n+1)^2\}_{n=0}^{\infty}.$$

3. Geometric patterns:

$$1,\ r,\ r^2,\ r^3,\ r^4,\ r^5,\ r^6,\ \ldots = \{r^n\}_{n=0}^{\infty}.$$

$$1,\ 2,\ 4,\ 8,\ 16,\ 32,\ 64,\ \ldots = \{2^n\}_{n=0}^{\infty}.$$

$$1,\ 3,\ 9,\ 27,\ 81,\ 243,\ 729,\ \ldots = \{3^n\}_{n=0}^{\infty}.$$

$$1,\ 4,\ 16,\ 64,\ 256,\ 1{,}024,\ 4{,}096,\ \ldots = \{4^n\}_{n=0}^{\infty}.$$

9.6 Alternating Patterns (Sequences)

1. Alternating linear patterns:

$$1,\ -2,\ 3,\ -4,\ 5,\ -6,\ 7,\ \ldots = \{(-1)^{n+1}\ n\}_{n=1}^{\infty}$$

$$-1,\ 2,\ -3,\ 4,\ -5,\ 6,\ -7,\ \ldots = \{(-1)^{n}\ n\}_{n=1}^{\infty}$$

$$1,\ -3,\ 5,\ -7,\ 9,\ -11,\ 13,\ \ldots = \{(-1)^{n}\ [2n+1]\}_{n=0}^{\infty}$$

$$-1,\ 3,\ -5,\ 7,\ -9,\ 11,\ -13,\ \ldots = \{(-1)^{n+1}\ [2n+1]\}_{n=0}^{\infty}.$$

2. Alternating quadratic patterns:

$$1,\ -4,\ 9,\ -16,\ 25,\ -36,\ 49,\ \ldots = \{(-1)^{n+1}\ n^2\}_{n=1}^{\infty}$$

$$-1,\ 4,\ -9,\ 16,\ -25,\ 36,\ -49,\ \ldots = \{(-1)^{n}\ n^2\}_{n=1}^{\infty}.$$

3. Alternating geometric patterns:

$$1,\ -r,\ r^2,\ -r^3,\ r^4,\ -r^5,\ r^6,\ \ldots = \{(-1)^{n}\ r^n\}_{n=0}^{\infty}$$

$$-1,\ r,\ -r^2,\ r^3,\ -r^4,\ r^5,\ -r^6,\ \ldots = \{(-1)^{n+1}\ r^n\}_{n=0}^{\infty}.$$

9.7 Summation Properties

1. Sigma notation:

$$a_1 + a_2 + a_3 + a_4 + \cdots + a_n = \sum_{i=1}^{n} a_i.$$

2. Addition of a constant:

$$\sum_{i=1}^{n} c = c \cdot n.$$

3. Distributive property of summations:

$$\sum_{i=1}^{n} [a_i \pm b_i] = \sum_{i=1}^{n} a_i \pm \sum_{i=1}^{n} b_i.$$

4. Alternating sums:

$$\sum_{i=1}^{n} (-1)^i a_i = -a_1 + a_2 - a_3 + a_4 - \cdots \pm a_n.$$

$$\sum_{i=1}^{n} (-1)^{i+1} a_i = a_1 - a_2 + a_3 - a_4 + \cdots \pm a_n.$$

9.8 Finite Summations

$$1 + 2 + 3 + 4 + 5 + 6 + \cdots + n = \textstyle\sum_{i=1}^{n} i = \frac{n[n+1]}{2}.$$

$$1 + 3 + 5 + 7 + 9 + 11 + \cdots + [2n-1] = \textstyle\sum_{i=1}^{n} (2i-1) = n^2.$$

$$1 + 4 + 9 + 16 + 25 + 36 + \cdots + n^2 = \textstyle\sum_{i=1}^{n} i^2 = \frac{n[n+1][2n+1]}{6}.$$

$$1\cdot 2 + 2\cdot 3 + 3\cdot 4 + 4\cdot 5 + \cdots + n\cdot[n+1] = \textstyle\sum_{i=1}^{n} i\cdot[i+1] = \frac{n[n+1][n+2]}{3}.$$

$$\tfrac{1}{1\cdot 2} + \tfrac{1}{2\cdot 3} + \tfrac{1}{3\cdot 4} + \tfrac{1}{4\cdot 5} + \cdots + \tfrac{1}{n\cdot[n+1]} = \textstyle\sum_{i=1}^{n} \frac{1}{i\cdot[i+1]} = \frac{n}{n+1}.$$

$$1 + r + r^2 + r^3 + r^4 + r^5 + \cdots + r^n = \textstyle\sum_{i=0}^{n} r^i = \frac{1-r^{n+1}}{1-r}.$$

$$1\cdot 2^0 + 2\cdot 2^1 + 3\cdot 2^2 + \cdots + n\cdot 2^{n-1} = \textstyle\sum_{i=1}^{n} i\cdot 2^{i-1} = [n-1]2^n + 1.$$

$$\tbinom{n}{0} + \tbinom{n}{1} + \tbinom{n}{2} + \cdots + \tbinom{n}{n-1} + \tbinom{n}{n} = \textstyle\sum_{i=0}^{n} \binom{n}{i} = 2^n.$$

9.9 Laws of Exponents

1. Sum of exponents:

$$x^n \cdot x^k = x^{n+k}.$$

2. Difference of exponents:

$$\frac{x^n}{x^k} = x^{n-k}.$$

3. Product of exponents:

$$[x^n]^k = x^{n \cdot k}.$$

4. Inverse of exponents:

$$\frac{1}{x^n} = x^{-n}.$$

5. Distribution of multiplications:

$$[x \cdot y]^n = x^n \cdot y^n.$$

6. Distribution of divisions:

$$\left[\frac{x}{y}\right]^n = \frac{x^n}{y^n}.$$

9.10 Factoring Methods

1. Difference of squares (conjugates):

$$x^2 - y^2 = (x - y)(x + y).$$

2. Difference of cubes:

$$x^3 - y^3 = (x - y)(x^2 + xy + y^2).$$

3. Sum of cubes:

$$x^3 + y^3 = (x + y)(x^2 - xy + y^2).$$

9.11 Binomial Expansion

1. $(x+y)^2 = x^2 + 2xy + y^2$.

2. $(x+y)^3 = x^3 + 3x^2y + 3xy^2 + y^3$.

3. $(x+y)^4 = x^4 + 4x^3y + 6x^2y^2 + 4xy^3 + y^4$.

4. For all $n \in \mathbb{N}$:

$$(x+y)^n = \sum_{i=0}^{n} \binom{n}{i} x^{n-i} y^i.$$

5. For all $n \in \mathbb{N}$:

$$(x-y)^n = \sum_{i=0}^{n} (-1)^i \binom{n}{i} x^{n-i} y^i.$$

Bibliography

O. Orlova, M. Radin, University level teaching styles with high school students and international teaching and learning. *International Scientific Conference "Society, Integration, Education"*, 2018.

O. Orlova, M. Radin, Balance between leading and following and international pedagogical innovations. *The proceedings of the International Scientific Conference "Society, Integration, Education"*, 1, 2019, 449–459.

M. Radin, V. Riashchenko, Effective Pedagogical Management as a road to successful international teaching and learning. "*Forum Scientiae Oeconomia*", 5(4) 2017, 71–84.

M. Radin, N. Shlat, Value orientations, emotional intelligence and international pedagogical innovations. *The proceedings of the International Scientific Conference "Society, Integration, Education"*, May 22–23, 2020, III, pp. 732–742. DOI: http://dx.doi.org/10.17770/sie2020vol2.4858.

Index

O

P

Q

R

CPSIA information can be obtained
at www.ICGtesting.com
Printed in the USA
JSHW031110060223
PP12303600001B/2

9 789811 261046